1
29.

ANCIENT WEAPONS IN BRITAIN

Helmets were rare in Britain until about the 9th century. However, a very few of extremely high quality, such as the Sutton Hoo example and the 7th C. Benty Grange one, mounted with a boar figure, have fortunately survived. To these we can add the outstanding Coppergate helmet found at Coppergate in York in 1982 and subsequently successfully conserved. This Anglo-Saxon example, with its elaborate, beautiful decoration and strong Christian connotations, was made in about 750 to 775 in Northumberland. (York Castle Museum)

ANCIENT
WEAPONS IN
BRITAIN

Logan Thompson

Pen & Sword
MILITARY

Dedication
To my late uncle, Bryan Bateman, for many years Curator of the Liverpool Museum Armouries.

First published in Great Britain in 2004 by
Pen & Sword Military
an imprint of
Pen & Sword Books Ltd
47 Church Street
Barnsley
South Yorkshire
S70 2AS

Copyright © Logan Thompson 2004

ISBN 1-84415-150-6

Typeset in Palatino 10pt

Printed and bound in Great Britain by CPI UK

Pen & Sword Books Ltd incorporates the Imprints of Pen & Sword Aviation, Pen & Sword Maritime, Pen & Sword Military, Wharncliffe Local History, Pen and Sword Select, Pen and Sword Military Classics and Leo Cooper.

For a complete list of Pen & Sword titles please contact
PEN & SWORD BOOKS LIMITED
47 Church Street, Barnsley, South Yorkshire, S70 2AS, England
E-mail: enquiries@pen-and-sword.co.uk
Website: www.pen-and-sword.co.uk

Contents

ACKNOWLEDGEMENTS

I am grateful to all who assisted in compiling this volume. Firstly, Angela C Evans, Curator of the Sutton Hoo collections and the Department of Prehistory and Europe at the British Museum, for her stimulating introduction to the Sutton Hoo Collection. To Barry Ager, also a curator of the Department of Prehistory and Europe, for giving me the opportunity to research Germanic Migration period swords; the rare privilege of detailed study of highly significant and complex Frankish throwing axes and, subsequently, a large seax collection. During these researches Barry periodically provided useful and thought-provoking suggestions on weapon problems and feasible technical advice. I am grateful to Virginia Smithson, departmental information officer, for providing an organized working base in the study room. I am also thankful to Ms Sovati Louden-Smith for her efficient, punctilious assistance in ordering numerous photographs.

For revealing comments on weapons in the prehistoric period, and for his very useful advice on arms, particularly on swords of the late Bronze and early Iron Ages, I am most grateful to Jonathan Cotton, Curator, Prehistory Department of Early London History and Collections at the Museum of London

I should like to thank Paul Hill, a most helpful and knowledgeable friend and colleague, with whom I undertook several joint weapon studies. These included Germanic Migration period swords from the pagan cemetery at Mitcham and a later, larger analysis of weapons and protective accoutrements recovered by Wessex Archaeology from the Saxon cemetery at Park Lane in Croydon. I am most grateful to Paul for his periodic assistance with this book, and especially for contributing the excellent Chapter 5, on Anglo-Saxon and Viking period spears.

I extend sincere thanks to John Eagle, the well-known military lecturer on the Roman army, for providing numerous accurate and, above all, realistic and exciting, illustrations featuring live action figures.

I am indebted to Elizabeth Hartley, Curator of Archaeology at the Yorkshire Museum, for allowing me to quote from the paper 'A Late Anglo-Saxon Sword from Gilling West, N. Yorkshire', which describes this significant weapon and its discovery. I am also most grateful to Professor John Hines, editor of Medieval Archaeology vol. XXX, 1986, published by the Society of Medieval Archaeology, for his kind permission to reproduce the fascinating technical drawings and to publish the informative blade paper written by Dr Brian J Gilmour.

Finally, I am particularly grateful to Professor Vera I Evison for permitting me to publish verbatim two sections from her enlightening paper 'A Sword from the Thames at Wallingford Bridge'. I am also indebted to her for granting permission to reproduce her drawing of 'the late Saxon sword from the Thames found in 1840'. The professor's article contributes markedly to a better and quicker understanding of the development and typology of English swords from the second half of the 9th century to the early 11th century. I strongly recommend her paper to those researching swords of this period.

PREFACE

This book describes the development and combat use of ancient weapons in Britain from prehistory to Hastings. Chapter 1 is a brief introductory study of the flint, copper, bronze and early iron weapons of prehistory. Thereafter, the work concentrates in great detail on weapons of the early Germanic period in Britain (a time for which, until recently, little in terms of weapon usage was understood) and continues with study of those of the Angles, Saxons, Franks, Vikings and the English during the 8th to the early 11th century. Examination of 11th century Norman weapons is included. The Viking chapter explains the reasons for the considerable increase of sword usage during this period, the complex Viking sword-hilt typology, and major technical sword-manufacturing changes.

I have undertaken detailed research since 1998 on 5th to late 9th century weapon collections in London and county museums. During this process I developed some new interpretations of Dark Age weapon handling and usage while attempting to formulate more plausible, wider theories concerning their use. I was compelled to adopt a new, special blade-measuring system due to the scarcity of surviving hilt components. This, rather surprisingly, actually proved most enlightening and useful. Until the works of Ewart Oakeshott and Hilda Ellis Davidson in the 1960s, and the papers of Vera I Evison, in particular her authorative work, 'A Sword from the Thames at Wallingford Bridge', little work had been done on the nature of weapon employment in antiquity. Much has been written on morphology and some work had been attempted on typologies. However, there is much to be said about the combat advantages and disadvantages of ancient arms and their performance in battle and I shall do so here. Where necessary this includes minor amendments to existing typologies.

It is difficult to be sure who wielded the first weapons in Britain, or what sort of arms they were, but it is true to say that, certainly by the neolithic period, warfare was widely employed, and evidence for it can be seen from excavations such as those at Crickley Hill and Cam Brae.

The former site, in Gloucestershire, was an undefended causeway enclosure that received an attack from an enemy armed predominantly with bows. The subsequent rebuilding of the site with a huge breastwork of ditch and rampart was clearly designed to keep such invaders out. The main point about modern excavations at this site is that they show warfare on a grand scale with defensive works and thousands of arrowheads within the perimeter. Crickley Hill was attacked at least twice in its long history. The indications are that whoever was doing the attacking was using missile weapons and was well organized.

It is, however, with the arrival in Britain of the first metal weapons that this book really starts. It becomes easier to trace the development of weapons from the dagger to the Bronze Age slashing swords, with their increasing emphasis on length and martial properties. During the Iron Age, the La Tene cultures influenced weapon design and function, and evidence from the period gives us a glimpse of how arms were handled and carried.

Model of legionary of about AD 100 with full kit less his dagger, reconstructed from finds from the Corbridge Roman site. Note the travelling cover on the scutum and his bronze mess tin. (English Heritage. Photo: S. Ellaway)

The Roman invasion ushered in a long period when highly disciplined, regimented, armoured soldiers with excellent weapons dominated Britain. The arms of this era contrast sharply with those used in Britain before the Roman conquest and, indeed, for many centuries after their departure. A chapter is thus devoted to sophisticated Roman weapons.

The book investigates the weapons used during the Migration period. Particular attention is devoted to the very effective large, double-edged, broad-bladed swords, which carried considerable status and symbolic significance. Ancillary weapons are also assessed. These include spears, javelins and shields. The various axe forms and their evolution are examined, including the Anglo-Scandinavian battleaxe – the terror weapon of the 11th century. The advantages and disadvantages of these weapons and their combat efficiency are analysed and the rarity of early helmets and body armour is highlighted.

Chapter 4 considers the wide range of Frankish arms produced from the 5th century onwards. These are included because they are periodically found in Britain and their design influenced some Viking axes, winged spear types and the Anglo-Saxon scramasax. Frankish weapons can also be seen in some museums in southern Britain.

Chapter 11 reviews the use of body armour, protective clothing and cavalry employed by the English and Norman French in the 11th century. It also examines the extent to which the Franks and the English strove, by passing a series of laws, to increase the availability of helmets, body armour and horses over a long period.

To better understand the nature of 11th century warfare and the effectiveness of all contemporary weapons, the final chapter is devoted to a detailed account of the Battle of Hastings.

Chapter One

Weapons of Prehistory

WEAPONS OF WOOD, FLINT, COPPER AND BRONZE

The first weapons, used for a very long time, were made of wood, and comprised clubs and spears. The effectiveness of spears was later improved by hardening the tips by holding them over gentle flames: a very early example of weapon technology development. An early missile was found at Schöningen, Lower Saxony in Germany. This is a long, spruce wood javelin, which is about 400,000 years old. It is 7.5ft (2.3m) long. To increase the range and power of the throwing spears they were subsequently sometimes launched with the aid of atlatls.

Flint

Later, tools and weapons successfully incorporated the natural substance flint, a hard silica found in the form of nodules. This has exactly the same hardness as steel (number 7 on the Mohs scale of hardness). Its use was a significant technical breakthrough in the production of various tools and arms in ancient times. After careful and skilful flint-knapping and polishing the substance was incorporated into arrow and spearheads, and then axes. The arrowheads of the neolithic period were either leaf-shaped and / or transverse (having one barb longer than the other and being slightly angled). In the Bronze Age, arrowheads were barbed and tanged, with a more triangular profile. Examples imported from Brittany were shaped in an almost perfect triangle. Axes used as tools, not weapons, were initially small hand-held ones but eventually became larger and had their head-shafts fitted into a wooden shaft, which was tightly bound with leather thongs. These could be used for cutting down trees. The sharp flint cutting-edges enabled much more efficient general-purpose implements and tools to be made. These included pointed knives, skin scrapers, hole borers and hoes. Because flint was only available in some areas it became a valuable barter-trade item. To obtain larger flint supplies, mining was undertaken at such places as Cissbury, near Worthing. The miners soon discovered that flints found underground were of better quality than those found on the surface. Other raw materials, such as quartzite, were also used for tool and weapon making, particularly in Scandinavia between 1800 and 1500 BC. 'The earliest of the fine flint daggers that the Scandinavians produced were long, slender, diamond shaped blades.'[1] Flint knives were eventually fashioned with blade and handle from one long piece of flint. The narrow section became a handle, which was bound with leather to provide a firm and comfortable hand grip.

Copper and Bronze

About 4000 BC, copper was first utilized for tool and weapon production in the

Middle East. This ore was initially cast in one-piece open clay moulds to fashion dagger blades and their handles, and axes and spearheads. Later, more sophisticated two-piece moulds were used. Swords originated from the realization that an extended dagger provided greater reach, which was more advantageous in combat. So the later daggers slowly developed into longer ones (dirks) and then eventually into rapier-style short swords. These were fashioned with double-edged blades. Unfortunately, copper ore had serious disadvantages. It was rather soft, causing blades to bend in battle, and was also difficult to hone to a sharp cutting edge. Such weapons were also inclined to snap. Consequently, sword lengths were necessarily short despite being reinforced with high, ribs.

About 3000 BC it was gradually appreciated that if another metal, such as tin or lead, was added to copper a harder alloy was created. This was called bronze. Among the many cultures who saw the advantage of mixing other metals with copper were the Sumerians. It was also soon understood that if the hot forging mass was tempered (cooled) quickly it provided a more malleable metal than when tempered slowly. This made possible the production of stronger swords, daggers and axes, with sharper and more resilient cutting edges, which were better in battle than copper ones. Furthermore, greater metal toughness enabled bronze arms to be made longer without the risk of them breaking or bending. If a sword included a complex and intricate hilt or pommel design, it was usual to add some lead into the metal mix because this cast more smoothly.

Early Iron Weapons

On very rare occasions in ancient times pure iron was also used in weapon manufacture. The sources were meteors, and those fortunate enough to discover them regarded the metal as invaluable. The Sumerian word for iron was 'heaven metal' and the Egyptians called it 'black copper from the skies'.[2] Such iron was being used in small quantities in Egypt about 3000 BC. Later, it was realized that iron could actually also be found in the earth, and if this ore was heated and then beaten, iron metal was produced. This discovery naturally increased the availability of iron supplies. However, metal created from such working was wrought iron, the edges of which could not be effectively sharpened. Eventually, after much experimentation, a more complex and successful iron-making method was invented. In this new process iron ore was heated with charcoal until the good metal was separated into a lump of iron called 'bloom'. The unusable remnants, which were full of impurities, were called 'slag'. The good iron was then beaten to remove any remaining impurities. This was then fashioned into bars, which were retained for the later manufacture of necessary tools, such as spades and sickles, and also swords. To produce the latter a long, narrow section was taken from a large iron bar. This was heated and fashioned, in a series of phases, to create the tang (handle section), the blade section and the point. The sword was next refined by hand-grinding on a stone. The final, perhaps most vital operation was the hardening, or tempering, process.

This involved putting the sword in an open furnace with charcoal, where it was left embedded in hot embers for about 7 to 8 hours before being tempered. This

Celtic iron daggers with bronze sheaths, probably made in Britain. From the River Thames at Hammersmith. Early Iron Age, 4th C. BC. (Museum of London)

process caused some carbon to be absorbed by the blade, which would have contributed to a hardening of the metal. This was an essential part of the casting process. When it was quenched the rapid temperature change caused the metal to be hardened still further, thus tempering the blade.

Smelting iron was first used extensively by the Hittites, about 1300 BC, in the Anatolia area of modern Turkey. This race was famous for their secret iron-weapon production, and carefully protected the new technological knowledge. However, about 1200 BC their empire was attacked by the Thraco-Phrygians. In the subsequent confusion some ironsmiths departed to other countries, where their secrets were doubtless divulged to others. The Iron Age had started.

Because iron was harder than bronze, and had sharper cutting edges, and its sources were now fairly abundant, iron weapons gradually became more common. They were more effective than bronze ones due to their greater strength. However, because iron-manufacturing secrets were still somewhat controlled, knowledge did not reach Europe until about 650 BC.

Celtic smiths were particularly accomplished metal craftsmen who produced high-quality artefacts in bronze and iron. 'Their metalwork created a great continental Iron Age tradition named after the Type site at Hallstatt in Austria'[3] between about 650 and 500 BC. The effects of their increasingly ingenious and successful metalworkings spread across northern and central Europe and south-east Britain. The quality of their arms decoration equalled that of their blades. Bronze scabbards and shields were enriched with confidently executed designs combining raised, ornate and stylized patterns sometimes further embellished with gold and enamel.

Research evidence perhaps suggests that, about 700 BC, a continental tribe made a military migration to Britain. These people were ruled by a military aristocracy using long, iron swords, or copies of these in bronze, and daggers and spears. They were often transported in chariots with bronze-cast wheel hubs. The chariots were drawn by ponies and, later, horses, with beautifully fashioned bronze snaffle-bits, cheek bars and bridle fittings. The race possessed martial competence, artistic ability, metalworking skills and advanced weaving competence. They spoke a language new to Britain.

It has been established that metalworking of weapons in Britain was pursued during the following periods: copper from about 2400 BC to about 2000 BC, tin bronze from about 2100 BC to about AD 100, and iron from about 700 BC until the arrival of the Romans. In Britain, bronze was of about 90 per cent copper, with some tin. The use of tin ensured that the alloy produced was harder, and also removed some of the free oxygen from the metal, producing a better casting. Elsewhere the tin was replaced by arsenic. This may not have been deliberate, but it was a product of the metal sources used. There are actually trace elements of arsenic in most bronzes.

Metalworking was undertaken in other countries much earlier than in Britain. For instance, copper was worked in the Chalcolithic period in Austria and northern Italy from c.2800 to 2500 BC; the Bronze Age extended from c.2500 to 800 BC, and the Iron Age from c.800 BC to AD 43 (the Roman invasion) and thereafter. These dates are the generally accepted eras of metallurgy. Metal Ages appeared earlier in the Near East and southern Iberia, where there were copper-mining settlements at Los Millares in southern Spain dating from the early part of the 3rd millennium BC.

SWORD DEVELOPMENT

Bronze Swords

The early evolution of the elegantly conceived slender bronze swords, better known as 'rapiers' because of their similarity to the much later 17th century duelling weapon, indicates they were initially made as thrusting weapons. 'The sword seems to have developed naturally from the knife of Minoan Crete and Celtic Britain at about the same time between 1500 to 1100 BC.'[4] Both these swords were of the thrusting type. Later, in the Middle Bronze period, heavier and more powerful, double-edged swords were introduced. This very different and heavier form was designed to achieve cutting and slashing blows. The difference in sword forms could indicate a change in fighting methods: the rapier may have usually been restricted to use in single-combat fencing; the later sword for slashing and cutting in close-quarter fighting. The heavy sword may also have been wielded by warriors on horseback.

It is appropriate to compare the two main sword types. Cutting ones were designed to achieve overarm hacking strokes directed mainly at the head, shoulders, and an opponent's shield. Thrusting patterns of narrower and lighter form were to make straight penetrative lunges at the body. Whilst employing the latter, the warrior was better protected by his shield than when making over-shoulder slashing strokes because the latter exposed, albeit momentarily, more of the body. The two blade forms were very different, as will be explained. By the late Bronze/early Iron Age some other sword forms were devised which could efficiently effect either cutting or thrusting strokes.

Thrusting Swords

The first Bronze Age blades were narrow and slender but rigid with long points. This particularly slight weapon is now called a 'rapier' and was designed to make straight penetrative thrusts to the body centre. The hilt (handle) was made separately from the blade and then attached to the fan-shaped blade top with metal rivets. The hilt grip was composed of wooden, or perhaps horn, plates. Such swords most probably had no guard and only sometimes a pommel, but blades were fashioned with pronounced medial ribs (see drawings on page 17). The joint between hilt and blade was an area of serious weakness, as the two sections were retained by too few rivets. Swordsmiths paid considerable attention to improving this vulnerable section. 'This was their weakness, for if lateral strains were placed on them there was little to prevent the rivets pulling sideways through the thin bronze.'[5] 'In nine cases out of ten the hilt rivets of the British rapiers have been wrenched sideways through the metal of the blades.'[6] These breakages were predominantly caused because the swords were too often used to effect horizontal slashing strokes at an opponent's face; a delivery for which the weapon was not designed. Eventually, by the middle Bronze Age (about 1100 BC) and after much experimentation, this serious difficulty was resolved by casting the blade and hilt in one piece. This discovery was of exceptional importance and significance because it made both thrusting and cutting weapons much stronger. Furthermore, the new blade and tang design was to continue with only minor changes throughout sword history. The tang was usually flat and was now an integral component with the blade. The actual hand grip of wood or horn was attached to the tang with rivets in a distinctive fan shape. Guards seem to have been non-existent.

Slashing and Cutting Swords

Double-edged blades in a new form, produced during the middle Bronze Age, were wider and thicker than hitherto and in a distinctive leaf or spearhead shape. They were still, generally, primarily intended to make hacking and cutting strokes, although the different pattern introduced about 900 BC could perform both cut and thrust strokes (see picture, right). Some blades are particularly leaf-shaped in form and thus appear somewhat rounded. Such large, heavy blades naturally added more weight and power to a hacking downward blow, so their effect was often more devastating. Swords were sometimes carried in finely fashioned bronze scabbards but the more utilitarian examples were carried in wood or leather ones. The length of such arms varied from about 17.72in (45cm) to 21.65in (55cm). Their form was usually attractive. Pommels, the important means of preventing a sword slipping out of the hand, retaining the grip on the tang, and providing at least some counter-weight to the blade, were not always used (see pictures on page 17). However, some early pommels were employed. Furthermore, swords used in Crete and Denmark in the early Bronze Age, and in the Rhone valley and Denmark in the late Bronze Age, did possess them. Some were a well-executed extension of the tang. By the late Bronze Age the swords had evolved into more modest shapes with slightly narrower blades which were somewhat more parallel in form to those of the middle Bronze Age period.

Holding a Sword

Whilst examining bronze swords I was regularly nonplussed by the small size of the hilt of many weapons. I initially assumed that perhaps the warriors who used them had slightly smaller hands than modern men, although this seemed unlikely from manuscript accounts of these warriors. Ewart Oakeshott, in his book *The Archaeology of Weapons*, explains the reason for the short hilts: 'One should not assume that a hand is too large because four fingers do not fit into the space between the pommel area and blade because the grip was only grasped with three fingers, the forefinger goes forward and below the shoulder whilst the thumb grips it fast on the other side.'[7] The swelling blade top does indeed provide additional grip tops for the forefinger and thumb. This grip mode would, nonetheless, make the sword hand vulnerable to opponents' sword cuts.

a   b   c

Three Bronze Age swords from Denmark. (a) Thrusting sword, c.1000 BC. (b) Cut-and-thrust sword with bronze hilt, c.900 BC. (c) Cutting sword with bone or horn hilt, c.850–700 BC. (National Museum, Copenhagen)

Sword Research

The Museum of London is an ideal venue for studying weapon evolution from the Stone Age to the late Saxon period. The Kingston-on-Thames museum also has an excellent collection of thrusting swords which were recovered from the River Thames. All were made in Britain and comprise four bronze alloy double-edged thrusting swords and one of the Carp's-Tongue type. The thrusting ones are all of the rapier type. Examination of these revealed the following:

Type	Length	Width	Approx. dates BC
1. Thrusting	16. 39in (41.5cm*)	1.3in (3.4cm)	1200–1100 Mid Bronze
2. Thrusting	21.06in (53.5cm)	1.38in (3.5cm)	1100–950 Mid Bronze
3. Thrusting	15.47in (39.3cm*)	1.34in (3.4cm)	1050–950 Late Bronze
4. Carp's-Tongue	21.45in (54.5cm)	2.6in (6.8cm)	750–450 Late Bronze
5. Thrusting	19.60in (49.8cm)	1.38in (3.5cm)	650–100 Early Iron Age

*All, or part of the hilt section is missing, therefore the swords were originally longer. The date of number 4 confirms the continuation of bronze sword manufacture into the early Iron Age. The presence of the very unusual Carp's-Tongue sword, albeit of modest form, with pronounced central rib and small built-in metal guard, is significant. Their strange title stems from the curious blade shape; its blade edges were very slightly tapering for most of the blade length, and then converged sharply to form the much narrower point. The lead content of this example is presumably high to guarantee good-quality casting of this rather intricate point (see picture on page 18). According to Ewart Oakeshott these swords were generally large, and a good example of one, also found in the Thames, is in the Brentford museum. The origin of these strange swords was probably northern France and elsewhere in Europe. Number 5 is a Hallstatt 'A' sword produced in the early Iron Age. This has about 40 nicks in the blade that were not caused by normal use; the damage was thus intentionally created, perhaps prior to the arm being cast on the water during a funeral ceremony. Swords 1 and 3 had lost their hilt sections, presumably for the reasons described under 'Thrusting Swords' on page 14. Number 3 is rather reminiscent of a Ballintober or Chelsea sword type, except that it has a flat blade section influenced by leaf-shaped blades, instead of the rather ovoidal shape of the true Ballintober ones. The width measurements shown here were taken at the widest points. These swords are collectively of special interest because the similarity of their measurements reflects disciplined standardization of manufacture.

It is appropriate now to briefly mention bronze spearheads. Many of these were fashioned in leaf-shaped forms with hollow socket ferrules into which the wooden spear shafts, often of ash, were inserted. Like the swords, these demonstrate the astonishing ability of contemporary smiths, who managed to produce weapons of delightfully proportioned elegance which were also markedly effective in combat.

The Iron Age

Iron tools and weapons were first used in the Hallstatt 'C' culture at the end of the Bronze Age. In 1931, C F C Hawkes divided this era into three periods: Iron Age A,

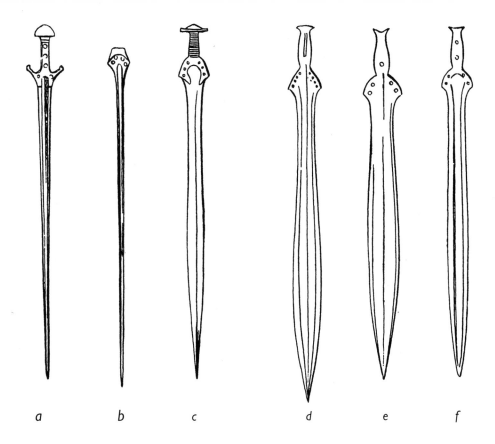

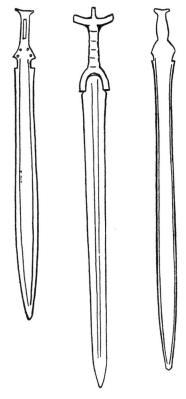

a   b   c   d   e   f

g   h   i

*Three Early Bronze Swords: a.
Crete, b. Ireland, c. Denmark.
Three Middle Bronze Swords: d.
England, e. Italy, f. Mycenae.
Three Late Bronze Swords: g.
Britain, h. Denmark, i. Austria
(Hallstatt). (E Oakeshott, from*
The Archaeology of Weapons,
Boydell & Brewer)

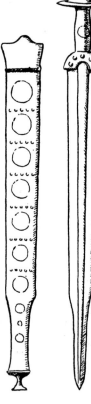

B and C. Iron Age A represented the Hallstatt immigrants and B and C the early and later La Tene immigrants. Iron Age B was considered particularly to have been the result of the Marnian invasions. Hallstatt is named after the Hallstatt site in Austria, La Tene after the Swiss site, and Marnian after the Marne district in France. This system has been continually modified and the old invasion theories have now been largely discounted, with people preferring to see the continental influence in the British Iron Age material culture coming in the form of small-scale settlement and the transferral of ideas and artistic styles, and not necessarily by the arrival of great hordes of warriors. This must surely be partially true. Doubtless there were some battles during this period. However, it seems likely that the primary desire of warriors was to present a fine display at all times to gain the maximum status and prestige. This was probably a significant cause for the production of such effective, highly attractive and artistic weapons, and diverse allied accoutrements fashioned by outstandingly capable contemporary smiths and weapon jewellers.

Hilt of 'Rhone Valley'-type sword. Late Bronze Age, from Switzerland, now in the British Museum.
(E Oakeshott, from The Archaeology of Weapons, *published Boydell & Brewer)*

Bronze Carp's-Tongue sword from the Seine, in the Musée de l'Armée, Paris.

'As with the greater part of our (British) Iron Age metalwork, the swords represent the aristocratic element in the Celtic-speaking society: they reflect on the one hand the barbarous inter-tribal strife of the warrior overlords – the equites of Caesar's three estates of the Celtic-speaking world – and on the other the assured craftsmanship and strange artistic genius of the armourers working under their patronage.'[8] The Iron Age in Britain is now generally seen as consisting of regional competing groups each influenced by Hallstatt or La Tene cultures to a differing degree. There was some migration, particularly from the Belgic areas of northern France in the 2nd century BC and from the Arras areas mentioned above.

Iron Swords

As soon as good-quality iron ore became more easily obtainable in adequate quantities and forging techniques became more sophisticated and efficient, iron swords were made in larger numbers. They gradually became the best available combat weapon because they were stronger than bronze ones and had sharper cutting edges. Initially, ownership of these was restricted to tribal chieftains and their personal retainers. Bronze weapons continued to be made for a very long time and these were traded with tribes still lacking iron-working ability. Conversely, trade of iron weapons was deliberately much restricted to ensure that tribes with the ability to make them maintained their military supremacy and domination over other tribes. The stronger, more reliable iron weapons were rather lighter and flatter in form than bronze ones, with longer, narrower tangs to which grips and attractive pommels were securely fitted. Bronze continued, of course, to be used to make some swords and their scabbards and chapes. Indeed, it was also used to make attractive sword furniture for iron weapons for a very long time indeed. This was because the alloy was better able to withstand corrosion as well as being an ideal alloy from which to fashion complex and extremely beautiful items such as scabbard components. From the 7th century BC the military elite started the tradition of carrying long, iron, single-handed swords, about 43 inches long, sometimes with a gold inlaid hilt typical of the early Hallstatt culture.

'These swords retain in their form most of the features of the earlier bronze types, but their main purpose was different. They were the long, slashing weapons of a chariot-using people. Their purpose is emphasized in many cases by the point not being a point at all, for it is of a rounded spatulate form, or cut off practically square, or drawn out into a sort of fish tail.' [9]

Swords used by Celtic-speaking peoples in Britain

A large number of swords and daggers discovered in Britain are of iron or steel, and their scabbards, chapes and hilt mounts are mostly bronze, although some are steel. This contrasts with the Continent, where the majority of scabbards and chapes are of iron. British scabbards have incised ornament along the full scabbard length as well as at the mouth, where the scabbard top snugly fits into the curved guard, and particularly on the chape.

'The distinctively British development (of swords) was to decrease the width of the sword blade and increase the importance of the chape as a vehicle for plastic ornament, and apart from the obviously exotic pieces in Group I (see below) the British series is almost from the first recognisably distinct from, even if it to some extent runs parallel to, contemporary continental trends.' [10]

Because it appears that the surviving sword evidence indicates that the weapons belong to the late La Tene I and the La Tene II types, 'it is unlikely that any of our examples go back beyond the middle of the 3rd century BC, and the time of the "Marnian" invasions heralding the establishment of the British versions of the La Tene culture.' [11] It is appropriate now to briefly examine relevant sections of Stuart Piggott's thorough typology on swords, daggers, scabbards, hilt types and chapes found in Britain. These embraced ten groups overall. Those relevant to us are:

Group I

This relates to fourteen daggers and their bronze sheaths almost certainly imported (or brought from) the Continent, probably of La Tene origin of the 2nd to 3rd century BC. The majority of these were recovered from the Thames.

Group II

Swords and scabbards of the Hunsbury type, of La Tene II derivation. These evidently represent the beginning of the main British series. Distributed from Somerset to Yorkshire, and probably produced mainly from the 2nd century BC to the early 1st century AD.

Group II A

Anthropoid-hilted daggers, associated with and broadly contemporary with Group III.

'With the establishment in England of the La Tene II type of long sword, in a scabbard normally of bronze, a fashion in weapons was set which was to endure in insular variants well into the 1st century AD and probably beyond. We cannot fix the lower chronological limits of our Group II swords, which probably continued to be manufactured, with comparatively little basic change in scabbard-design, until the Roman conquest – a restricted group of swords and scabbards derived from the continental La Tene III type [my Group V] appear probably as a result of the Belgic invasions but do not seem to have influenced the native design in the south.'[12]

The latter can be generally classified as swords and daggers of the Bugthorpe type: a north-eastern British development from Group II of the 1st century BC which probably continued into the 1st century AD.

Group IV

Swords and scabbards of the Brigantian type, which developed in north Britain from Group III from early in the 1st century AD to about AD 100.

Group IV A and IV B

The hilt guards can be divided into cocked-hat hilts and crown hilts.

Group V

Swords and scabbards of the Battersea type, not developed from insular forms but representing the continental La Tene III form, and probably Belgic, early in the 1st century AD. It should be noted that: 'in southern England, as we have seen, it is not possible to detect any evolution from Group II swords and scabbards parallel to the developments in the north east.'[13]

It seems the majority of surviving swords and their scabbards were made from just after 150 BC. The manufacturing process continued thereafter until the Roman conquest (AD 43) in southern Britain and in the north of the country, and in Scotland for much longer. The majority of the arms, according to Stuart Piggott, belonged to the Hunsbury type – the main British series. All these are distinctively insular and characteristically British. They are particularly interesting because their construction indicates that Celtic-speaking armourers and swordsmiths were, presumably, able to acquire from within Britain supplies of good-quality iron ore, copper and tin.

'The iron swords of Group II have narrow blades, not usually more than 1.5 inches (4cm) wide at hilt guard and up to over 2.5 feet long usually with a slight median rib or thickening. Hilt and pommel were of perishable material such as wood, bone or horn which in no instance has survived, but a thin ogee-arched bronze or iron hilt guard is usually present, fitted to the scabbard mouth, and surviving iron tangs are from 4 to 5 inches long (10 to 12.5cm) sometimes burred over the top.'[14]

The reduction of the blade width was a highly significant design change instituted, perhaps, on account of a change in the purpose of sword use. Doubtless its previous aim of delivering hacking and slashing strokes was to a degree retained but perhaps now it could also be used to deliver straight thrusting lunges.

Chapter Two

The Roman Army and its Weapons

Before examining the weapons and armour used by the Romans in Britain we should understand something about the highly efficient army that used them:

'For centuries, the expansion and preservation of the Roman Empire rested upon the broad shoulders and discipline of heavy infantry legionaries and due to the efforts of these carefully recruited, ruthless, highly trained and well led Italian soldiers that Roman civilisation spread and developed. It was the legionaries who ultimately ensured efficient, uniform administrative standards, gigantic ambitious building projects including an extensive network of metalled roads which guaranteed rapid, controlled movement of large armies, mesmeric forum oratory, a guaranteed peace and real certainty of its continuation for the majority and luxurious life-style for the few. Without the powerful army, one of the best in history, and certainly the most effective for the longest period, none of the former would have been possible.'[1]

THE ROMAN ARMY

Military Organization

The army, possibly the best in the world in about 60 BC, later became even more dominating and successful. By the time of Claudius's invasion of Britain in AD 43 it was considerably more formidable because the Roman government now had the administration to properly support large military operations. It was a highly efficient killing machine. The well-armed force comprised two closely allied but different organizations: the legions and auxiliaries. The legions were predominantly elite, armoured heavy-infantrymen, and at the invasion of Britain most of these were Italian. Such men served in a legion numbering about 5500. Apart from some cavalrymen used as scouts and messengers, those involved with artillery and engineer duties, and specialists such as doctors, engineers and surveyors, the remainder were all infantrymen. These highly trained soldiers served 25 years. A legion was divided into 10 cohorts, 9 of which comprised 6 centuries, each of 80 men. A century was commanded by a centurion. The other cohort had 5 centuries, each of 160 veteran troops. Most cohorts thus had a strength of 480 men. The smallest sub-unit was a section of 8 men commanded by a corporal who lived in one barrack room and shared a tent on campaigns. Very often, detached forts were garrisoned by a cohort. After Britain had been conquered, three legions were permanently camped in the province, at York, Caerleon and Chester.

About half of the army comprised auxiliaries. 'Under the Emperor Augustus these non-Roman soldiers were organised into regular army units, in the main with Roman officers. Many retained national differences of equipment and mode of fighting.'[2] These were arranged into infantry units such as a cohort (480 men). The cavalry served in units of about 520 troopers. Sometime a cohort was a mix of cavalry and infantry, which made it very flexible in small-scale battles. There were various other useful specialist auxiliary units, such as archers and highly trained river-crossing troops. The cavalry units were of the greatest importance to the Romans as they had not developed their own large mounted arm. Their important battle roles were: firstly, to encircle enemy foot-soldiers who were retreating, and secondly, to protect their own infantry when it was put under pressure by enemy

Roman officer and soldier in Britain about AD 20. Note officer's unusual but authentic helmet and elaborate metal cuirass and shoulder plates. (John Eagle)

foot-soldiers. Auxiliary units, with the exception perhaps of cavalry, were less well equipped than the legions and received inferior pay and rations. However, such units certainly proved themselves very valuable in numerous campaigns, and the Romans benefited greatly from their help. Such units included Balearic slingers, Hamian archers with their deadly recurved composite bows, and Batavian river-crossing experts. A unit of the latter gained honour during the invasion of Britain in AD 43 by being the first troops, during an opposed crossing of the River Medway, to reach the enemy bank by swimming across, and then surprised the Britains by attacking the charioteers' horses. This success was repeated soon after, when they successfully swam across the River Thames during another opposed river-crossing. In the summer of AD 82 or 83 the Roman general Agricola fought a major battle in north-east Scotland (at Mons Graupius) against some 30,000 Caledonians under the command of Calgacus. Agricola possessed 8000 auxiliary infantry and 5000 auxiliary cavalry in addition to his legions.

Unconventionally, he successfully launched his auxiliary infantry against the Caledonian front line and then partly encircled their army with cavalry, eventually driving them from the battlefield with losses of about 10,000 men. This indicates the faith Agricola rightly placed in the ability of his auxiliaries and their combat capability.

Soldier Training and Discipline

The Roman army devoted much time to tough and highly disciplined training, which was planned to closely resemble real combat, for recruits and experienced soldiers alike. Recruits received concentrated training, which took up all of their time for about four months. This was overseen by a centurion who rewarded slack behaviour with smart beatings with his olive-wood cane, and was supervised on-ground by sergeants and junior NCOs. Constant drill was the most regular means of ensuring that recruits moved smartly, correctly and with united cohesion when in a squad. Its most important purpose was to produce an automatic reaction to any order. They became used to wearing thick-soled hobnailed boots, short-skirted leather tunic, armoured breastplate, and helmet with its neck guard and cheek plates. They were taught how to carry their large shield and weapons when on the march. Legion uniforms were surprisingly well standardized, possibly due to their mass production at military-supervised ordnance centres. Auxiliary clothing varied from one unit to another.

All infantry units undertook regular route marches several times a month. This was because the ability to march long distances fast and then immediately fight a major battle was a prerequisite to deterring neighbouring states from causing trouble to the Empire. During marches, soldiers carried 50 to 60 pounds of equipment at 4 miles per hour, thus covering 20 miles in 5 hours. If longer distances had to be covered quickly they implemented a forced march pace of four and a half miles per hour. Considerable weapon training was undertaken. Javelin (pilum) throwing was constantly practised to ensure range and accuracy and then the ability of a unit to deliver a dense co-ordinated mass of missile weapons. Sword use with dummy wooden arms used against an opponent eventually led to the employment of real swords with points covered. Sword handling against wooden posts accustomed the recruit to the stunning effect of the blade against a firm object.

Roman army discipline was extremely severe:

> 'The loss of rank, seniority, or privileges were frequent punishments. Dishonourable discharge meant losing both the shelter of the unit and more important, the grant of land or citizenship on retirement. Men found guilty of barrack room theft or desertion could be, in military law, beaten to death by their comrades. Cowardice in battle sometimes led to decimation – literally, the execution of every tenth man in the unit concerned; the survivors were put on a ration of barley.'[3]

Doubtless, the multitude of minor indiscretions of army life, such as being late on parade, wearing poorly cleaned equipment, or insubordination, were dealt with by a firm beating, extra cleaning or guard duties, or loss of minor privileges.

ROMAN UNIFORMS AND WEAPONS

Clothing

At the time of Claudius's invasion, legionaries wore a short-skirted leather tunic and a pair of leggings. On their feet they wore thick-soled hobnailed boots (caligula)

Legionaries of the 14th Legion with pila and scuta form defensive rank. (John Eagle)

rather like heavy sandals. Over their sleeved tunic was a cuirass of iron hoops protecting the shoulders and trunk and breastplate and backplate. However, many soldiers were still equipped with a mail cuirass that stretched to below the knees, providing protection against cuts and thrusts. This form of protection continued in some units until the end of the 1st century. The head was protected by a metal helmet (the coolus) with neck guard, a ridge at the front and cheek pieces. The only protection below the waist was an apron of metal-bound thongs suspended from the belt (cingulum) protecting the groin. Centurions wore shoes, a leather corslet, shaped to fit the body, and bronze greaves to guard the shins. They carried their sword and dagger on the opposite side to the legionaries, that is, on the left-hand side.

The auxiliaries were far less standardized in their dress than legion soldiers. Consequently their appearance varied considerably from one unit to another, and somewhat reflected their regional origins. Helmets were simpler than those of the legionaries and they often wore mail under their leather tunics. Hamian archers from Syria wore conical helmets with cheek guards of a Middle Eastern type.

Weapons

The weapons described here are those of about AD 20 and would probably still have been in use at the time of the Claudian invasion. In this section I am indebted to John Warry's *Warfare in the Classical World*. The legionary had four weapons. These were two javelins (pila), which were used in varying forms from about the 4th century BC until the end of the 3rd century AD, when they were replaced by a longer spear thought to be more efficient against cavalry. One pilum was heavy; the other light, with the longest about 7 feet. They had slender, iron heads set on wooden shafts. These were often made with a soft, metal shaft below the head, which was designed to bend when it struck an opponent's shield. Both were very effective in attack or defence at up to about 30 yards because when launched they could either kill an adversary or penetrate his shield. In the latter case, the pilum head bent preventing it from being withdrawn from the shield and thrown back. Additionally, with the pilum attached to the shield it became an encumbrance and was often discarded reducing the warrior's personal protection. 'As time went by the heavy pilum got smaller and the light one larger, until in the 1st century AD they were identical. By AD 100 the pilum had shrunk further, and a bronze weight was added to maintain armour piercing capability.'[4] At this period, however, the legionaries would approach an enemy, launch their javelins, then close, with their opponents employing their heavy shields as another weapon. These were semi-cylindrical in shape, and made of leather-covered laminated wood bound with bronze strips.[5] They had an iron or bronze boss in the centre of the outer face and metal emblems that may have been unit markings. The projecting shield boss would be smashed hard into the face of an enemy in order to push the man back in disorder or knock him off his feet. At this point the legionaries employed their third weapon, the gladius sword, with a 20-inch parallel-sided blade and ivory hilt. The Romans always fought in close-packed shoulder-to-shoulder ranks, so the swords were used in a series of short, upward stabbing strokes to kill adversaries.

They realized it was unnecessary to thrust their blades in deeply to slay a man because such action wasted time in recovering the weapon. The combination of these actions, skilfully employed, caused the legions to achieve many victories, particularly against ill-disciplined barbarians.

The final weapon, the dagger (pugio), with its long, pointed blade, was carried in a separate scabbard supported by another cross belt.

'The grip was of wood, ivory, or bone and the leather scabbard often intricately tooled. During the military reforms of Marius (104–101 BC) creating a more professional Roman army, the weapon was an optional extra. They were used as an emergency, last resort arm and for administrative tasks. When the sword form was changed in about AD 70 to one of a parallel-sided blade and sharp point the pugio was unmodified. Soldiers were encouraged to adorn the hilts and scabbards in a smart and distinctive manner to inculcate personal and regimental pride.'[6]

A new dagger form was introduced in the 3rd century AD.

Javelins were often exceptionally effective in combat. This was particularly so at the conflict which ended the Boudiccan revolt in about AD 60 or 61 which took place in the Midlands, possibly at Mancetter or Towester. The Roman force under General Paullinus was only about 10,000 men. These were confronted and vastly outnumbered by the host of over-confident British tribesmen led by Boudicca. The Romans formed up on slightly higher ground in a defile protected by a forest at their rear. Infantry deployed in close order in the centre with auxiliaries and cavalry on each wing. The British deployed in rather loose array. Behind them were numerous wagons loaded with their supportive families and children, who anticipated another great victory. The Romans then eventually taunted and tempted the Britains to make a frontal

A high-status Roman gladius sword with hardened blade cutting edges and ornate 'tinned' scabbard, found in the River Thames at Fulham. Blade length: 22 1/8in long (63.4cm). (© Copyright the British Museum)

Roman soldiers of the late Empire, 3rd to 4th C. Note white-painted shield depicting Christian symbols, and dismounted trooper with cavalry sword and metal pagan badges attached to cross belt. (John Eagle)

Right soldier, typical of those in Britain during the 1st C. AD; left, Roman regular soldier of 4th C., the last of those to defend Britain. (John Eagle)

attack up the slope and, at the critical moment, when they were about 30 yards away, discharged their javelins in two concerted barrages.

This dense fusillade involved a total of about 15,000 missiles, causing numerous casualties and abruptly halting the barbarian advance. The Romans then drew their swords and at once advanced in a closely formed, disciplined manner, rapidly killing any wounded or incapacitated Britains they found. They drove back the tribesmen, penning them against their own wagons. The battle then deteriorated into a massacre in which both women and tribesmen were slaughtered without mercy. The number of British dead was calculated at the exaggerated figure of 85,000 while the Romans lost only about 400 men. The primary cause of this emphatic and rapid victory was the disciplined and effective use of the pilum.

Auxiliaries' Weapons

These troops, with the general exception of specialists and elite cavalry units, were less well armed than the legionaries. Many carried a spear or two javelins, the spartha sword and large, oval shields. When the sword pattern was changed they were issued to the auxiliaries. Troops of Celtic origin used the flat, Celtic-type shields. Specialist troops naturally carried their special arms, such as heavy bows, heavy javelins, slings and relevant accoutrements. They wore mail of the ringed or scale types. As previously mentioned, cavalry was vital to Roman commanders, and their uniform included a helmet with cruciform reinforcements on the skull. A trooper also wore his mail or scale armour while carrying the flat, oval shield. At this point, stirrups had not been invented, but special saddles prevented the man being unhorsed and ensured his weapons could be effectively employed while he was mounted. His weapons were:

> 'a long sword derived from a Celtic type and a light spear (lancea) suitable for thrusting overarm (Roman cavalry did not normally couch their spears under arm). The historian Josephus also mentions a quiver of darts (light javelins) attached to the saddle. This is confirmed by Arrian who describes cavalry exercises in which up to twenty darts were discharged in one run.' [7]

The swords used by Roman auxiliary troopers were long, slender, acutely pointed blades with a stiff section like a flattened diamond.

British Military Opposition

The Britains lacked an organized and disciplined army. Instead there were only 'the levies of tribes who might or might not join a common cause. Lengthy campaigns were more or less impossible for them, since the vast majority of the troops were farmers.'[8] Perhaps the most regular military organization was the charioteers who, when deployed simultaneously, could provide a total force of several thousand war chariots. Their owners formed the aristocracy who, because they were not bound to the land, had plenty of time to practise martial skills. These they had obviously perfected because Julius Caesar was most impressed by them, mentioning that charioteers sometimes ran out along the chariot pole and stood on the yoke. Chariots were usually driven at a full gallop and were highly manoeuvrable. The fear inspired by the noise of chariot wheels, galloping horses and flying javelins was sufficient to instil disruption in barbarian armies; but not in the Romans. However, they presented a novel threat to the Romans, who treated them with respect. Chariots were neither large nor heavy and could not be employed to break up massed infantry ranks. The wheels, which were 3 feet in diameter, were not fitted with scythe blades, but they were well designed, with a specially sprung seating platform between them. Their metalwork, particularly of the wheels, illustrates clearly the advanced competence of Celtic iron and bronze-smiths. They were drawn by two ponies and had a two-man crew: the driver sat, whilst the standing warrior, protected by his Celtic shield, hurled javelins. Sometimes they would stop, allowing warriors to fight on foot. If outnumbered they would then remount the chariot and withdraw. 'One of their favourite stratagems was the feigned retreat, to draw off small parties of the enemy who could then be tackled by the chariot-borne troops leaping down to fight hand-to-hand.'[9]

Little seems known of barbarian cavalry. Possibly, horsemen were regarded as members of a lower aristocratic order. It is feasible that their primary military purpose was to charge directly at an enemy while rapidly hurling a series of javelins: a disconcerting and uncomfortable experience for the enemy. This seems to have been the device employed by the Brigantes, a powerful tribe in northern Britain. Perhaps they were actually more successful in battle than surviving accounts indicate. 'Celtic cavalry served both Caesar and Aulus Plautius extremely well and it seems likely that British horsemen were very similar. The probability is that they too belonged to an upper class that could practise regularly and perhaps supply their own horses.'[10]

Foot-soldiers who were farmers could only serve during certain times of the year. Consequently, they were not organized into small military units and received no basic on-ground training. Therefore, in battle they fought with spears, stripped to the waist with painted bodies, in an ill-disciplined, milling mass. Doubtless, tribal chieftains, accompanied by their personal guards, equipped with some protective mail and helmets and carrying long Celtic swords and shields, could put up a fairly proficient combat performance. Their swords were indeed potentially powerful arms but they needed space to wield effectively in attack and defence. This dictated fighting in open order, which put them at a disadvantage and meant that they were locally outnumbered when fighting Romans in their tightly packed, disciplined ranks. We now appreciate that most British warriors were less well equipped, trained and battle-experienced than Romans. Courage and elan were insufficient to counter disciplined troops used to defeating barbarians.

Artillery

When the Romans invaded Britain they took a considerable number of artillery pieces. As was their custom with any new technology, they copied the artillery achievements of others. 'Constant refinements and technical improvements in Hellenistic times led to significant increases in the range and power of catapults.'[11] However, the Romans subsequently made their own refinements to the novel weapons. Each legion had an establishment of fifty guns, which could put down a considerable barrage. These were particularly daunting against massed infantry and cavalry. Additionally they were even more useful when employed against the gigantic and potentially formidable hill forts in south-west Britain. To deal with these they used massed batteries of ballistas firing darts of about 27 inches long at up to 400 yards with great accuracy. These were aimed at wall tops to keep down defenders' heads and at the buildings within the fortifications. All the Iron Age forts which refused to surrender before hostilities began were successfully captured with the use of artillery. The devastating and conclusive effects of Roman artillery are revealed by archaeology. Skeletons have been found with darts embedded in their spines, and skulls found with tell-tale small, square dart holes. The Britains were, not surprisingly, disconcerted by, and fearful of, the Romans' lethal use of artillery against them. Furthermore, because their own missile weapons comprised only javelins, sling stones and arrows their warriors were usually significantly out-ranged by artillery.

Roman artillery basically comprised heavy guns such as the onagers, which fired very large stones, and the lighter, more accurate ones like the ballista (stone thrower), scorpio (dart firer) and later the cheiro ballistra (dart firer). The latter two fired forms of large heavy bolts, darts and arrows. The various launcher types thus provided a versatile choice of guns to employ against different targets. 'The Romans, for their part, devised a mobile unit known as the carroballista cart catapult of which there is representation on the Trajan column. It was drawn by horses, and could be operated by its crew of eleven while the animals remained harnessed to it.'[12] It should be noted that it was unnecessary for the Romans to employ the enormous mobile ones in Britain, but very-slow-moving siege towers armed with battering rams and light artillery were employed against large, well-protected stone-built cities in the eastern Mediterranean.

> 'The ballista looked rather like an overgrown crossbow mounted on a stand, but in fact the ballista worked in a slightly different way from a crossbow. Instead of a bow, the ballista had a pair of arms joined by a bowstring. The arms were held in two upright bundles of rope or sinew which acted as large torsion springs and created a huge force.'[13]

The loading and firing process entailed the operator pulling back the string using a winch mechanism to pull against the great force created by the bundles of rope, until the string was caught by a catch. The operator then loaded the weapon. 'When the lever was pulled the catch was released and the string and bolt flew forward at high speed.'[14]

From these descriptions of the disciplined ability of Roman army personnel and their versatile range of excellent weapons it is easy to appreciate the capability of the Roman army. They eventually conquered the whole province of Britain up to Hadrian's Wall. Thereafter, they retained control for about 350 years upholding the maintenance of sophisticated government which inculcated widespread growth of civilization and a higher quality of life that would not otherwise have been possible. Above all, the army made possible, for the most part, a very long era of peace.

Chapter Three

Germanic Swords of the Migration Period

In the late 4th century Germanic peoples settled in southern Britain. This initial, small migration was probably deliberately fostered by the Roman government of Britain to allow them to recruit warriors and naval assistance in return for land and payments. Such practice had been well established in the Roman Empire for about two centuries. Indeed, the usurper Allectus, Britain's ruler from AD 293 to 296, maintained a large contingent of loyal Frankish mercenaries. 'That the Roman policy of employing mercenaries was continued in the post-Roman period is well illustrated by the history of the semi-legendary Vortigern (a Romano-British leader) and Hengist and Horsa [Germanic military leaders].'[1] The warriors involved were called foederati. Their leaders were periodically presented by the Romans with elaborate chip-carved belt furniture of cast bronze, sometimes decorated with silver inlay, as symbols of their allegiance to the Romans. 'Such fittings were Roman military issue in the late 4th century, and are found above all in German graves including those of Saxons who had retired home to die.'[2] However, these belts were not purely military, because they appear to have been issued also to officials in the civil administration. Several have been found in the Germanic settlement at Mucking on the Thames Estuary.

The Germanic soldier-farmers and their families initially mingled peacefully with local British populations and served their new masters for a long time. The Germanic tribes represented among the migrants comprised Angles, Saxons, Jutes, Franks and some Alamanni. In this chapter these are usually collectively referred to as Germanic peoples because their specific deployment is still somewhat unclear. However, it has been established that Jutes settled in Kent and the Isle of Wight.

Later, a major disagreement with the Romano-British government led to a serious revolt. According to Robert Jackson in *Dark Age Britain* this occurred about AD 442, when Hengist, the Saxon leader, rebelled against the government on account of its long-term failure to deliver promised supplies. This successful uprising resulted in the conquest of areas of east and south-east Britain. Subsequently, large additional numbers of other Germanic peoples moved to Britain with the new, more aggressive aims of conquest and permanent settlement. The Germanic migrant groups, although not a unified people, possessed similarities in material culture, language, social structure, pagan religions and the employment by their warrior elite of long, double-edged, often broad-bladed, single-handed swords. This chapter examines these and assesses their distinctive features, production methods, social and symbolic significance and combat use and effectiveness. The study spans the period from about AD 400 to 800.

Chip-carved belt furniture of cast bronze from the Mucking settlement. Possibly made in late 5th C. Britain and presented to a senior foederatus. (© Copyright the British Museum)

GERMANIC SWORDS

Significance

Contemporary swords were rare, and their ownership was restricted to kings, war-band leaders and their elite personal guards. Possession of one indicated high social class and status. Study has been devoted to calculating and debating the numbers of swords simultaneously in use during the Migration period. To a degree, totals can be assessed by adding the number of confirmed recoveries from cemeteries to the number discovered at battle sites, rivers, gravel pits and other remote locations. However, these will obviously span different periods. My studies lead me to believe the total number employed at any one time was small for two reasons. Firstly, they were extremely expensive to make, and secondly they were only used by the nobility and elite warriors, whose numbers were very small in proportion to the overall population. The rarity of weapons is to a degree confirmed by several cemetery excavations. At Kingston in Kent only two swords were recovered from 308 graves; at Hollywell Row one sword in a 100 graves, and none at all in the 123 graves at Burwell. Conversely, at Sarre in Kent a sword was found in every 10 graves. It is particularly encouraging, however, that swords are still being discovered periodically. A recent example was during the excavation of the early Saxon cemetery at Park Lane in Croydon.

Early in our period, the weapons of the non-military upper class (civilians who were freemen and thus entitled to carry arms) who settled in Britain were remarkable for their crude simplicity in sharp contrast to swords. The weapons comprised spears, often of poor design and quality, rather small, round wooden shields and simple daggers. The famous scramasax dagger, initially developed by the Franks as a seax arm, did not appear in Britain until about the late 7th century.

Swords always indicated that their owner was of high status and a most capable warrior. Warriors were probably selected for training as fighting men at the age of fourteen or younger. The best of these eventually became the contemporary 'regular soldiers'. The sword, used in conjunction with the shield, was the most devastating, feared and serviceable weapon on the infantry battlefield, capable of cutting through spears, shields, padded leather jerkins and even the rare hauberks (mail coats). They frequently inflicted fatal wounds to heads, shoulders, arms and legs. This has been regularly authenticated by grave remains such as those from Eccles in Kent investigated by S J Wenham in 1989. Furthermore, their decisive use could bring about a battle's climax by, for example, killing the opposing commander. Sometimes this was sufficient to achieve victory because badly disciplined second-line troops with uncertain morale might well flee the battlefield. Naturally the chieftain's personal guard would probably continue fighting, even to the death, to avenge their lord's demise.

Some examples were considered to achieve greater prestige and power with age and the war-like reputations of previous owners. 'A sword which had belonged to King Offa was still in the possession of the Royal family of Wessex two hundred years later, for it is mentioned in the will of Prince AEthelstan.'[3] It is possible that other important swords were used for considerable periods: 'The repeated evidence of the sagas about swords being handed on is too clear to be passed over.'[4] Certainly they were generally preserved and well cared for, as indicated in the scabbard section on page 48. Pagan Germanic peoples believed the swords possessed magical and supernatural powers, and to enhance this the upper scabbard plates, scabbards and chapes (but not the blades) were sometimes inscribed. These included various symbols, such as swastikas and zigzag and wavy lines and, very occasionally, runes. Ellis Davidson, in her book *The Sword in Anglo-Saxon England*, suggests that the swastika had either religious or magical significance for the Anglo-Saxons and that it may have been connected with the worship of the god Thunor. 'The zigzag lines may also have some special meaning: they suggest a lightning symbol, and there are other devices on pommels of this period which seem to be symbolic rather than purely ornamental.'[5] Sometimes swords were named, like Beowulf's 'Naegling'. 'Any evidence which archaeology can offer concerning the naming of weapons is of great value, for while a large number of sword names, nearly two hundred in all, have been preserved in literary sources, these are relatively late.'[6]

To our pagan forebears, swords therefore represented military power, social status, and a magic talisman that had reassuring associations with their heathen religious beliefs, which sometimes encouraged warlike attempts to emulate the martial exploits of their war gods. Their pagan religion greatly encouraged warriors to undertake combat with enthusiasm and determination.

Military Situation in Britain after AD 442

The military situation in Britain during the 5th and early 6th centuries, after Hengist's rebellion in the south-east, was discussed by the monk Gildas in his book *On the Ruin and Conquest of Britain*, written in the 540s. He described how the British regrouped with some vigour: 'God strengthened the survivors, and our unhappy countrymen gathered around from all parts, as eagerly as bees rush to the hive when a storm threatens.' Such resolution is not surprising. During their long occupation of Britain the Romans had established excellent administrative, defensive and communication systems. Furthermore, the experienced rulers of the civitas capitals doubtless helped to stiffen citizens' nerves and morale while encouraging the people to suppress the rising. Additionally the population would surely have included many capable ex-Roman officials and soldiers with useful military knowledge who contributed to the reorganization of the defence forces. The Romano-British evidently possessed one important military advantage over their pagan enemies. They maintained small but disciplined cavalry units under the command of experienced officers. Possibly the troopers used long, straight swords similar to those described by Ewart Oakeshott in *The Archaeology of Weapons*. He states: 'The swords used by auxiliary cavalry had long, slender, acutely-pointed blades with a stiff section like a flattened diamond.'[7] Cavalry was probably organized, equipped and trained on the old Roman army system, and thus was potentially effective against Germanic foot-soldiers incapable of creating a defence, in the open, against mounted shock power.

Damage had been caused during Hengist's rebellion, although its extent has not yet been precisely established. However,

'The British defences were organized under a man of Roman race named
Ambrosius Aurelianus, whose descendants in the second generation were
still ruling somewhere in Britain when Gildas was writing. For a time there
was a struggle between the Britains and the new invading forces but it was
ended by a British victory at a place not now to be identified, called Mons
Badonicus.'[8]

This occurred in about AD 495. Robert Jackson in *Dark Age Britain* convincingly
suggests that the Mount Badon battle site was probably at Liddington Castle
adjacent to the present village of Badbury. This was a formidable Iron Age fort
near Swindon which had been reoccupied by the Romano-British. The British
commander is reputed to have been the legendary and fictional Arthur. It is
important to state that no firm historical evidence exists to prove that Arthur ever
existed. Nonetheless, if the Britains were not commanded by Arthur they would
very probably have been commanded by another British general. Such an officer
was likely to have been titled a Dux Bellorum, a military commander specifically
of cavalry. Robert Jackson suggests that the British commander, with a force of
about a thousand cavalry, was penned up in the fort by a much larger host of about
three to four thousand Saxon infantry. They were, perhaps, attempting to conquer
a large, new territory. Then, as he says, 'The cavalry, eventually spotting some
enemy weakness, launched a devastating charge that broke the attacking army and
put it to flight, inflicting enormous slaughter on it during the pursuit that
followed.' On reflection, this achievement is not surprising, as the great majority
of the Saxon host comprised many second-line (civilian) troops.

This important British success prohibited further westward Germanic
encroachment for the next thirty to forty years. Furthermore, it is likely the victory
discouraged to a significant extent fresh Germanic immigration while persuading
some existing migrants to return to their continental homelands.

Sword Production

'Your Fraternity has chosen for us swords capable even of cutting through
armour, which I prize more for their iron than for the gold upon them ... The
central part of their blade, so cunningly hollowed, appears to be grained with
tiny snakes ... Such swords by their beauty might be deemed the work of
Vulcan.' A letter of Cassiodorus, Secretary of Theoderic the Great, King of the Ostrogoths
(AD 441–526), expressing the King's thanks for the receipt of a gift of several swords from
another Dark Age king.

Swords were made from iron, contemporaneously a valuable raw material whose
quality varied considerably from one region to another. High-grade pure ores often
contained greater quantities of the beneficial substances manganese and titanium
and far fewer impure substances such as phosphorus and sulphur. Therefore it was
more efficient to process good ores than poor ones because they required less time
and effort to be spent hammering out the unwanted impurities, and moreover, the
blades produced were of better quality. Roman mines noted for producing good-
quality ores included those at Noricum on the Danube and those of the Sana valley
in Bosnia. 'Noricum ores gave excellent results because they contained an

unusually high proportion of manganese and titanium whilst being relatively free of common impurities.'[9] One question hovers over Germanic early broad-bladed sword research in Britain: that of the number of swords actually made outside Britain and the extent of those produced in Britain. It seems likely that in the early Migration period most were manufactured on the Continent and carried by early migrants to Britain. Possibly, some leaders brought swordsmiths with them who may have then set about sword production. This supposition assumes they either found local ore sources or were supplied with iron blanks from abroad. The current problem is that it is not usually possible to positively trace their source of such iron in Britain.

Sword Blade Types and their Characteristics

The typology of Migration swords was first established by the Swedish scholar Elis Behmer, who categorized them into nine groups. Later, Ewart Oakeshott produced a simplified version of this concentrating on only four groups. The most common sword type in Britain during the early Migration period was Behmer's Type IV, which was used mainly from the mid-5th to the 7th century. These have a broad, two-edged blade with the cutting edges approximately parallel up to just above the spatulate point section. The latest comprehensive research on this subject, which supersedes the work of both Behmer and Oakeshott, was by Wilfried Menghin, who produced a far more detailed categorization. His work *Das Schwert im Frühen Mittel-alter* details high-quality continental swords and their associated grave goods.

I have classified some 40 single-handed Migration period swords, which include two main forms: the large, broad-bladed, parallel-sided types and the narrower, shorter and slightly tapering types with more-pronounced points. The latter group comprises two sub-groups, one of which contains slightly shorter and narrower swords than the other. At the Salisbury & South Wiltshire museum, the average measurements of six broad-bladed swords were: overall length 35.03 inches, blade length 30.76 inches, and blade width just below the guard 2.03 inches. The rule-of-thumb measurements of such a sword type are: overall length of up to 36 inches if the pommel survives, blade length 31–32 inches, and blade width of up to a maximum of 2.20 inches. An elegant, beautifully fashioned, slightly tapering sword at the Croydon museum was 32.95 inches long (without pommel), with a blade length of 28.20 inches and a blade width of 1.79 inches. It should be noted that the dimensions of Migration period swords generally differ slightly. Tapering blades are most generally narrower at the guard point than parallel-sided ones. According to Ellis Davidson, in *The Sword in Anglo-Saxon England*, the Type IV sword measurements were: blade lengths from about 34 to 37 inches, and blade widths from about 1.8 to 2.18 inches. These equate very closely to my research. 'This form of sword was inherited from the prehistory Iron Age of central and western Europe.'[10]

The Swords in Combat

The single-handed, parallel-sided, double-edged weapons were primarily designed to deliver very powerful blows from above the shoulder. They were always used in conjunction with a round shield carried on the left arm (see Chapter 6). The blade's point of balance was low and it was this area of the blade that warriors attempted to direct at a pre-selected portion of an adversary to achieve the maximum force. These arms generated the most powerful and effective hacking and cutting strokes. Some examples are particularly substantial with overall lengths of more than 37 inches. A particularly magnificent and intimidating one is on display at the Devizes museum. The swords could also deliver slicing blows at an opponent's legs. They may occasionally have been used to deliver straight thrusts, although the long, broad blade would have made these difficult to achieve. However, despite delivering effective above-shoulder hacks, they harboured some disadvantages apart from their limited stroke range.

Above-shoulder blows could momentarily leave the right side of the body unprotected; the swords were rather clumsy and unwieldy, and the tiny pommel was insufficient to act as a counter-weight to the blade weight. The low point of balance meant that sword recovery between above-shoulder strokes soon became tiring for the warrior. I can confirm this, from practical weapon handling. A further disadvantage was that the swords were difficult to manoeuvre into a defensive position.

The elite soldiers who carried these rare weapons had been trained from the age of about fourteen to become highly combat-efficient warriors. This gave such recruits plenty of time to become accustomed to their use. They inflicted horrific and, doubtless, fatal injuries, particularly on the left side of opponents' heads, shoulders, arms and legs. Instances occurred in which a single blow had cut through a warrior's hauberk, shoulder blades and ribs down to the waist. King Theoderic dealt such a blow to Odovacar, King of the Ostrogoths, in 493, which split his body from shoulder to waist.[11] It is probable that veteran warriors delivered their powerful strokes with very careful and slow deliberation to ensure every strike was accurate and effective.

The possibility that employment of these swords caused fatigue prompted me to consider they were carried only by physically powerful men. This theory was partly supported during research of the Germanic swords from Mitcham pagan cemetery. Two graves, numbered 65 and 89, containing broad-bladed swords, were occupied by men of 6ft 2in and 6ft 1in respectively.[12] Also Härke's suggestion that the skeletons exhumed from weapons burial sites in pagan Saxon cemeteries in England indicated taller men than those from non-weapons burial sites is perhaps unsurprising.

Tapering Swords

Contemporary thinking about Migration period swords indicates that the majority used in Britain were of the parallel-sided type examined above. It is perhaps conceivable that some tapering ones were brought to Britain by early migrants. However, my research at various locations indicates that there were far more tapering ones than I anticipated, which somewhat contradicts this theory. Of the 40 Migration period swords (see table below) I classified, 19 were of the large parallel-sided type, 18 of the two tapering types, one short Frankish sword (found in a Saxon grave) and an unusual one with a broad blade of 2.5 inches, possibly Alamannic, reminiscent of Behmer's Type III. So at present the numbers of the two

Important swords from Petersfinger, Wiltshire.
Left: Sword 52 of about AD 600 with scabbard
plate with horizontal decoration. Cocked-hat
pommel with curving sides and deep, crescentic
flutes. Very slightly tapering blade. Accompanied
by dark green translucent glass bead probably
used as an amulet. Overall length: 34.75in,
(88.5cm); blade length: 31.20in (78.6cm).
Right: A particularly interesting sword (60) of
the 7th C., with significant pommel of
pyramidal shape decorated with lenticular
flutes and silver plate. A simple foliate motive
decorates the pommel front. The two gilded
pins, one of which is fashioned as a bird or
serpent head, would have retained the
'sandwich' pommel section, now missing.
However, a shallow concave slot on one side of
the pommel may have facilitated a ring. If so,
this weapon would have been a most
important example of a 'ring' one.

Its upper scabbard plate is
distinctively embossed in a
late Roman manner in the
style of the earlier Frankish
period, marked by the Gallo-
Roman motives. (Salisbury &
South Wiltshire Museum)

types seen are currently very similar, but of course, future research may well radically alter the data deductions. It is also possible that the tapering swords examined were actually made later than the parallel-sided ones for the reasons suggested below. Unfortunately, owing to some 1400 to 1500 years of corrosion the dating differential between the two forms is often very difficult to establish. This is also aggravated by the absence of additional datable material.

Tapering swords, with their narrower blades and more acute points, were naturally lighter and less unwieldy and therefore somewhat easier to handle. Consequently they were more versatile and useful in combat as they could be employed in several ways. The sword-stroke range included straight thrusts at the central body, horizontal slashing ones at the face and legs, and hacking blows launched from above the shoulder. These blows were probably less fatiguing to make than a similar stroke using a heavier arm. Perhaps the sword's most significant benefit was the way they could more quickly and effectively be moved to a defensive posture. These characteristics may explain their popularity. With my colleague Paul Hill, who examined many examples of this sword type, I eventually decided that their numerousness could have been due to warriors' appreciation of their advantages over the clumsier parallel-bladed form. Such fighters must therefore have persuaded swordsmiths in about the late 5th and early 6th centuries to produce blades of this form. To summarize, I consider the vast majority of swords brought to Britain in the late 5th and early 6th century were of the parallel-sided type. Thereafter, the narrower, tapering form were gradually introduced and became, by the end of the 6th century and early 7th century, much more popular.

MIGRATION PERIOD SWORD BLADE STUDY

LOCATION	NO. OF SWORDS	TAPERING	TYPE IV	MISC	REMARKS
Mitcham	7	4	3	-	
Salisbury	6	4	2	-	
Devizes	6	1	4	1	Transitional
Croydon	6	4	2	-	
British Museum	8	2	5	1*	*Short Frankish Sword
British Museum	2	-	2*	-	*Long narrow parallel-sided blades
Museum of London	4	3	1*	-	*1 Alamannic (parallel)
TOTALS	39	18	19	2	

Pattern-Welding

The best sword blades in this period were strong and flexible with considerable springiness and attractive surface patterns. All were made by the very expensive and complex pattern-welding system. In this technical and specialist section I have enlisted the aid of some sections of the research paper, 'Swords of the Saxon and Viking Periods in the British Museum: a Radiographic Study' by the acknowledged experts Janet Lang and Barry Ager at the British Museum. They state:

> 'The term "pattern-welded" is used to describe a process of welding together twisted rods to build up pattern-welded blanks (bars) from which swords, daggers, and spear heads were later made. Examples of these blanks (bars)

Three Migration swords from Mitcham pagan cemetery. Left: single-handed tapering sword designed for both hacking and thrusting strokes. A late 6th C., pattern-welded example bearing scabbard remains on blade. Blade length: 30.91in (785cm). Centre: another, similar retaining early small pommel but missing its upper scabbard plate. Note tang, which seems rather narrow for a weapon of 34.84in long (88.5cm). Right: a very robust and intimidating 6th C. Migration sword of Elis Behmer's Type IV. Cutting edges are parallel for most of the blade length. Pommel is more sophisticated, being fitted with two copper alloy gilded pins which originally retained the 'sandwich' components to a thin bar. This powerful hacking weapon is 35.39in (89.9cm) long. (Kingston Museum and Heritage Service)

2 ft.

1 ft.

0 inches

can be seen at both the British Museum and the Museum of London. Pattern-welding is characterised by the presence of patterns on the blade which were originally visible to the eye, although in corroded specimens a radiograph may be needed to reveal them.

'The technique of pattern-welding, with which the present paper is mainly concerned, is known to have been used from the 3rd century AD and appears to have evolved from piled structures made by the Celts and Romans (Lang 1984). These piled structures were constructed by forge-welding together a number of sheets or strips of iron laid on top of each other. The composite structure was then forged to the required shape. The components were frequently arranged with layers of carbon-rich iron alternating with layers of low carbon iron.'[13]

Later, in the 3rd century AD, the twisting became more complex, and true pattern-welding could be said to have started. In this process, a complex structure was built up by plaiting or twisting iron strips (which were sometimes themselves made up from a number of strips) and then welding them together (Anstee & Beck 1961). The patterned piece was used to make the central sections of a sword blade, or part of a knife or spearhead. Some of the earliest pattern-welded swords, from the 3rd and 4th centuries AD, were found at Nydam in Schleswig Holstein (Schurmann 1959), in the Rhineland, at South Shields (Rosenquist 1967), and at Canterbury (Webster 1982). 'The fairly wide carbonised cutting edges were welded later to the central edges of the weapons.'

During the early Migration period high-quality iron ore supplies from such places as Noricum on the Danube and the Sana valley in Bosnia became disrupted owing to the disintegration of the Roman Empire. This compelled many smiths to use poor-quality bog ores, or to rework old iron-production centres. This encouraged them to use the pattern-welding production system. 'They adopted it in their efforts to deal with impure bog ores which formed their main source of iron.'[14]

I have mentioned the beautiful blade patterns that emerged, as if by magic, during the final blade-cleaning process after the blade grinding and polishing. These appear not only on the blade but also naturally within the blade structure. The internal pattern-welded form obviously influenced to a degree the external blade pattern. Pattern-welding can take various forms: chevron, herringbone, straight or curving. The beautiful, swirly, serpent-like outer-blade pattern raises the question of whether pattern-welding was primarily employed as a means of achieving a most attractive blade or excellent practical weapon. Although direct evidence is rather sparse it seems most likely that the Germanic warrior elite selected the arm because

Pattern-welded Migration sword of 7th to 8th C. Crooked guard and triangular pommel, both comprising thin sections of iron, brass and horn. Surviving traces of linear and knot-work decoration. Owing to corrosion, the blade is now in flat form, enabling the pattern-welded chevron pattern to be clearly seen. (Salisbury & South Wiltshire Museum)

of its combat-effectiveness. After all, their lives depended on it, but, doubtless, they were delighted with the status that its appearance bestowed on them. Theoderic the Great's pragmatic view of such swords was probably shared by most warriors.

The Lang and Ager study involved the radiographing in the British Museum research laboratory of 142 swords that had been discovered in the British Isles. This revealed that the overall proportion of weapons in the collection made by pattern-welding was actually rather high. Their interesting deductions relate to study of this particular collection only:

1. 64 per cent of the Anglo-Saxon and Viking swords were pattern-welded.

2. The proportion of pattern-welded arms rose dramatically after about AD 500 then fell again during the 9th to 10th centuries.

3. All the 7th century examples were pattern-welded.

4. No sword could be found belonging to the 8th century. Presumably this was due to changed burial practices.

5. A small proportion of swords had pattern-welded inscriptions on both blade faces but the remainder were without pattern-welded decoration of any kind. These were restricted to the period after AD 800 (see Chapter 6).

6. It is of interest that later inscribed sword blades from England were not pattern-welded although the inserted letters were pattern-welded. This seems to be true of most similar swords described in other studies and is further evidence for the decorative rather than functional purpose of pattern-welding.

I deduced from point 2 that to some extent, at least, the fewer pattern-welded swords discovered in early pagan cemeteries (classified as such after research) the older the non-pattern-welded swords, and the settlement in which they were found, are liable to be. This supposition can, to an extent, be confirmed by the Hill–Thompson study of the Mitcham Saxon Migration cemetery, where non-pattern-welded swords were in the majority and could probably have been dated earlier than we initially calculated.

Hardening and Tempering

Good-quality iron blades could be made even harder and more flexible by a quenching and tempering process. Ellis Davidson explains: 'the blade was first heated then dipped into a liquid. This cooled the metal very quickly. It was then reheated to a correct, lower temperature which a competent smith could detect from the colour and cooled again either rapidly or fairly slowly.'[15] It was a very skilled procedure because if the blade was cooled too fast the metal tended to be over-hard and less flexible. The flexibility could, however, be restored without much loss of hardness by reheating the blade then repeating the tempering process. Much experimentation was undertaken during ancient times with various fluids to discover the best one for quenching. Water was an obvious and easily available one but sadly not efficient 'since it boils at a low temperature and forms a steam barrier to the passage of heat, and in modern processes liquids such as oil or molten lead are used.'[16] Experiments were made with differing compounds such as vegetable oil, honey, moist clay and urine.

Blade Fullers

Some but possibly not all swords of the early Migration period had broad, shallow fullers (troughs) down both blade sides extending to just above the point section. These had nothing to do with blood runs. They were designed to lighten the blade, increase sword flexibility and concentrate blade weight towards the cutting edges. Some weapons, not in this study, such as early Danish ones of the Roman Iron Age, possessed fullers comprising several deeper, narrow grooves. It may be that early in the Migration period only some swords were fullered.

Unfortunately, this point is often difficult to determine because on the large numbers of badly corroded arms even with a radiograph the feature cannot easily be detected. Being so shallow they were rapidly corroded and cannot be detected by the eye, even with slanting sunlight, or by X-ray. Therefore, one cannot conclude that a blade was not fullered, only that it is not possible to determine one way or another: an inevitable but unsatisfactory conclusion. However, research has provided some useful insight on this subject. A sword in good condition with fullers discovered in the British Museum (recovered from the Thames) was identical to an example in poor condition from the Mitcham Saxon cemetery which lacked a fuller. Therefore the latter would initially have had fullers. This curiously fortuitous incident suggests that more Migration swords were fullered than available evidence suggests. It also endorses the fact that weapons deposited in rivers are likely to be less corroded than those from the earth.

Xeroradiograph of pattern-welded 9th C. Anglo-Saxon sword from Hurbuck. It shows central blade core without cutting edges. This made from a lower layer of three rods and top layer of three rods, each rod comprising several iron strips which were first twisted together. Note chevron pattern resulting from manufacturing process. (Copyright © the British Museum, with thanks to Royal Marsden Hospital)

Sword Components

We now appreciate that during the Migration period in Britain the general quality of pattern-welded sword blades was high or very high. We should not forget, of course, that some less sophisticated and cheaper blades of 'steely iron' (not made by the pattern-welding process) were also of good quality, assuming they had been quenched and tempered. These blade achievements thus often contrast markedly with the utilitarian and rather crude construction of many hilt items such as pommels and guards. There also seems to be a sparsity of well-designed and artistically decorated sword fittings. There were, of course, some exceptions in Britain, such as the Sutton Hoo, Cumberland and Crundale hilts. The Cumberland sword has several 7th century gold mounts on the horn hilt, but there is usually a marked contrast between very-high-quality blades and their inferior sword furniture. This tradition continued for a very long period in both imported and, possibly, locally produced blades for much of the Migration period. Doubtless there was at that time a scarcity in Britain of highly skilled hilt jewellers capable of making artistic, decorated gold hilts.

Sword furniture was fitted to the tang and scabbard. The items were the main guard, tang grip, upper and lower guards and pommel, all fitted to the tang. The chape, upper scabbard plate, and strap and belt fittings were mounted on the scabbard.

The Tang

This is the top extension of the blade, which emerges from the rather square blade-top shoulders in a narrow, flat and slightly triangular form. It was necessary to cover the rather sharp tang point to protect the palm of the sword-hand. On some swords tangs seem rather modest in ratio to the blade length, width and weight. However, considerable effort was necessarily devoted by swordsmiths to fashioning tangs because this weapon section had to be sufficiently resilient and flexible to absorb the major shock effects created by blades striking an opponent's weapon, shield or body without snapping. The tang was, of course, often constructed from the pattern-welded mass during the forging process, which would beneficially strengthen the section. The blade shoulder tops were generally rather square, but the fashioning of this area does vary considerably. Some appear rounded (perhaps owing to corrosion, which is often particularly marked in this area) whilst others were more precisely fashioned, suggesting their creation by a high-calibre smith. The sharp tang tip was necessarily covered with a small iron welding to protect the palm. Eventually, this area was covered by a short, small flat iron bar (see pommel section below).

The Pommel

The pommel, sited at the tang end, could help to prevent the sword slipping out of the back of the hand. Sometimes the tang point fitted directly into a little pommel but usually these were attached to the little iron plate, or bar, later known as the 'upper pommel bar'. Early Migration pommels were usually very small, being little more than a slightly raised metal section, but later they became larger, fashioned as a small, bronze cap, often gilded. They were attached to the tang cover bar by spot-welding. The little pommels were the diminutive beginnings of the long cocked-hat pommel tradition (see Appendix A on page 50). Eventually, the cap ends were attached by gilded bronze pins to the upper pommel bar. Later pommels were elaborated and enlarged with the introduction of a lower pommel bar. Between this and the upper pommel bar was created a 'sandwich' of alternate layers of horn, wood or bone tightly fitted together. These were also attached with bronze pins. This naturally increased pommel's size and strength. The hilt area became more sophisticated when the main guard was matched in design with the pommel. The introduction of two bronze plates into the main guard, between which was also a 'sandwich' of horn, wood and bone (matching the pommel), made the guard stronger. A 'sandwich' might also include more valuable metals such as brass. On higher-status swords, bronze pommels and their pins were sometimes silvered and decorated (see picture opposite). Because the hilt was easily seen it was an ideal place to display rank and status with elaborate and expensive decoration. Pommels are a useful sword-dating aid; without them this important fact can be more difficult to establish.

Ring Pommels

During the 6th and 7th centuries some rare and highly significant swords were fitted with decorated rings attached to the ends of the pommel. Several have been found in Kent (see picture opposite). On early patterns one ring was loose within the other. Ellis Davidson convincingly discusses this subject: 'these loose rings vary in size and are all elaborately made with bands of ornament which usually matches decoration elsewhere on the hilt.' Later types, also fixed to the pommel top side were fashioned in a different manner with one ring permanently attached within the other. These were called 'ring knobs'. Davidson further intimates their possible purpose: 'they possessed ritual and symbolic significance and swords bearing them are regarded as special. They may have been intended to commemorate the gift of a sword from one warrior to another to symbolise their alliance as blood brothers.' They might also, very possibly, have formed an intimate function in the important ceremony during which a king took the oath of allegiance from a man selected to membership of his Royal bodyguard. She finally comments 'the swords upon which rings were fixed were owned by kings and leaders, the men who gave out swords to their followers.'

Grip Construction

This is the hilt section which enclosed the tang providing a hand grip, often of horn or wood plates. Possibly, treated horn was the most popular method. Sometimes, however, the tang may have been completely enclosed in a carved wooden sleeve that was drilled down its centre. Into this the tang point was inserted and then tapped down from the rear until it reached the guard. It would have provided a most secure grip. I first noted the possibility of such a method when working on a Frankish seax, the tang sides and edges of which had been enclosed with wood. There were remnants of a wooden sleeve. The prerequisites for grips were very secure attachment to the tang, provision of firm and comfortable handhold, and guarantee that it would not shift, or turn, when the sword struck heavy blows. Grips were sometimes partitioned into three or four ridges to provide a better handhold. This practice establishes a slender link with the Roman spartha hilt grips, which had several indentations. Such a style affords a reassuring grip for the hand and fingers. At the end of the Migration period some grips were securely fitted onto a traverse bar sited below the lower pommel guard and into a second bar below attached to the guard.

Guards

My detailed research on Migration swords revealed very little evidence of sword guards. This is particularly interesting when compared with the often-intricate guard systems of much later swords. In a very few cases there was evidence of a former guard in the form of faint, rectangular, narrow staining on the blade top below the tang. This area tends to be much lighter in colour than the blade below it. The absence of guards is because they were constructed from perishable materials such as horn, bone and possibly hardwood, which, over the centuries, have completely disintegrated. However, from the blade staining we can deduce they were straight and narrow. It could be assumed from the small size of guards that their purpose was not considered significant by Migration period warriors. During classification of the four swords from the Croydon Saxon cemetery I was fortunate enough to discover a fragment of the original guard on one of them. This was half an inch long and just over a quarter of an inch wide. It matched the rectangular guard marks across the blade and positively confirms that early guards were of modest form. Subsequent analysis of the fragment showed that it was horn. Analysis of deposits on the

Rare Kentish silver-gilt hilted sword of very early 6th C., recovered from large ironbound coffin in grave 204 at Finglesham, Kent. Estimated grave date about AD 525. Owner possibly a Jute aristocrat buried with sword, silver studded shield, spear and knife. Note free-running (early pattern) ring and beautiful pommel with their fine decoration. (Picture reproduced with kind permission of The Institute of Archaeology, University of Oxford, and comments from: The Anglo-Saxons edited by James Campbell, published by Penguin Books)

same sword, particularly along the tang top edge, also proved to be horn, indicating that the whole hilt had been covered with plates of this substance. The theoretical purpose of the guard was to prevent the hand sliding down onto the sharp blade edges and being cut. Even though it is clear that the warrior's primary defence was the shield, it is still rather surprising that the hand was not better protected. Although I have no evidence to support my theory, I think warriors may possibly have worn on their sword hand a leather gauntlet with its upper side made of thick, boiled leather.

Scabbards and their Attachments

From the remains of fairly complete scabbards of Migration swords and major fragments of others it appears that their construction was often thorough and sophisticated. This is rather surprising when one considers their somewhat simpler forms in later times, and indicates a sensible determination to preserve and protect their precious blades. Some comprised a rather complex series of leather-bound wooden laths. The scabbard was cushioned with an inner lining of sheep wool intended to secure the blade a little more firmly and which provided the added benefit of rust prevention by virtue of its lanolin content. Traces of such wool are sometimes noted on sword blades, in addition to fragments of leather and wood. Next to both sword-blade sides were thin and short wood laths slightly narrower than the blade. 'These innermost pieces of wood may have been intended to serve as stiffeners to the scabbard, but it is equally likely that they are guides to prevent the point of the sword catching in the leather parts of the sheath.'[17] These laths were tightly bound with thin leather and then enclosed in another thicker wooden sheath that was bound with robust leather. To provide extra protection the outer leather was sometimes untreated, thus retaining animal hair as an additional barrier against rain. To protect the scabbard base a round bronze or iron chape, often comprising a narrow U-shaped strip was fitted (see picture on page 124). This ensured the lower leather-and-lath end sections were firmly secured. More elaborate bronze alloy chapes covering the complete bottom section of the scabbard were occasionally used. Sometimes, the chape edge was extended upwards on one scabbard side in a J-shape on the scabbard front edge. This ensured better sword protection when moving through high foliage.

The scabbard mouth was to a very small degree safeguarded by an upper scabbard mount or plate. These are usually a modest, gilded bronze alloy rectangular strip encircling the scabbard and attached at the rear by two bronze pins. The practical value of this feature in protecting the scabbard mouth was minimal, so it can best be regarded as a purely decorative, perhaps vestigial, prestigious feature. On special, rarer swords the scabbard mount can be in a much larger and more conspicuous form. A splendid example can be seen fitted to a large sword in the Devizes museum. This is a large and deep mount of moulded bronze alloy (originally gilded) created in a beautifully crafted perforated form. It has sophisticated and intricate decoration depicting two zoomorphic animals (possibly fierce dogs) with heads tilted backwards over their bodies, and is embellished with other unusual decorative features. This sword has a blade width of 2.30in wide beneath the guard. It must surely have belonged to a war-band leader, or possibly a little king. Its zoomorphic decoration indicates it came from north Germany. Upper scabbard mounts were simply decorated in various patterns. It is of particular interest that a Migration sword in the Croydon museum (M1992) retains a gilded copper alloy upper scabbard mount decorated with incised lines in keeping with the Mitcham-Kempston type outlined by Menghin, two classic examples of which exist on two of the swords from the Mitcham cemetery now in the Kingston museum (Hill and Thompson), and one in the Devizes museum. It seems possible that these were made by the same smith, who may, perhaps, have travelled to various settlements selling his expertise to warriors.

Swords were carried either on the sword belt or in a cross belt providing a baldric slung over the right shoulder. This incorporated a frog in which the sword was retained. In the former method decorated bronze strips were attached vertically to the top outer scabbard edges through which the waist belt passed. A bronze strap-end attachment was fitted to the lower rear scabbard edge. To this was fitted another strap which extended to the back of the waist belt, where it was retained. This fulfilled the important function of lifting the lower scabbard backwards retaining it at an angle which kept it above undergrowth, reduced the chance of the warrior falling over the blade, and made it easier for him to draw his weapon. The waist belt front provided two areas for displaying status in the form of decorative mounts.

We have studied the substantial swords carried by the warrior elite of the Germanic Migratory tribes that settled in Britain. These were of three types: broad-bladed swords with parallel sides, broad but tapering ones, and narrower-bladed and tapering ones. Many blades, particularly in the 6th and 7th centuries, were made by pattern-welding. These were strong, flexible and particularly effective when swung from above the shoulder. The carbon hardened-steel edges were extremely sharp and penetrative. We have examined how the clumsy parallel-sided swords, perhaps the most devastating of the period, were gradually replaced by tapering ones, which made a much wider range of strokes possible. Examples with the intriguing and mysterious serpent-form blade surfaces were both formidable and beautiful, and were highly prized and much sought after. We noted that, in contrast to the many very good blades, the quality of the sword furniture was generally utilitarian. This contrasted with the very high standards of these artefacts sometimes achieved on the Continent. During the Migration period the sword was the best infantry weapon in northern Europe. Possession of a rare, high-quality, pattern-welded sword indicated high status. More importantly, it provided warriors with the best practical means of dominating the battlefield and defeating their foes while protecting themselves. The weapon was also revered as a magical religious talisman and was believed to provide links with their pagan gods.

There is an important final point to appreciate about Migration swords. Today, many on display may be very worn, corroded and have few remaining sword furniture embellishments. In museum showcases they can appear rather lifeless, inanimate and somehow unexciting. Surviving blades are at times very thin as a result of extreme metal loss, which is only to be expected after what may be 1500 years of corrosion. Originally they were naturally heavier and much more robust. Nonetheless, anyone who actually handled well-preserved examples, particularly of the parallel-sided category, would be impressed by their intimidating size and potential power. Furthermore, when carried daily by Migration warriors they were immediately recognizable as formidable and magnificent arms. Beautiful modern reproduction swords created by contemporary swordsmiths enable us to appreciate how truly impressive and fearsome such arms appeared in ancient times.

APPENDIX A

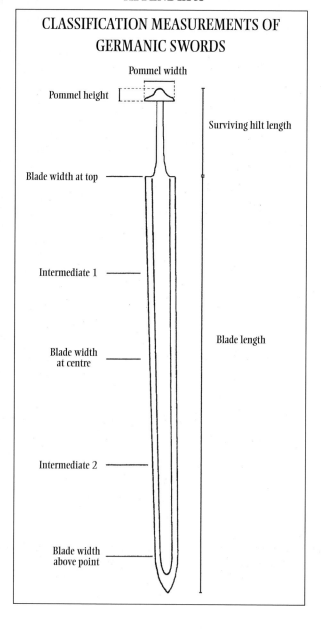

CLASSIFICATION MEASUREMENTS OF GERMANIC SWORDS

Pommel width

Pommel height

Surviving hilt length

Blade width at top

Intermediate 1

Blade length

Blade width at centre

Intermediate 2

Blade width above point

Chapter Four

Influence of the Franks

'They [the Franks] are a tall race, clad in close fitting garments with a belt encircling the waist. They hurl their axes and cast their spears with great force, never missing their aim. They manage their shields with great skill, rushing on their enemy so fast that they seem to fly more rapidly than their javelins.' *Sidonius Appolinaris (430–483)*.

This chapter covers the period from the 5th to mid-8th century and, in particular, the Franks, a powerful continental Germanic people who became notable under the emperor Charlemagne (742–814). We shall briefly examine their weapons because some have been found in Britain and others greatly influenced arms employed by other Germanic tribes. For centuries the Franks were known for having armies that were predominantly infantry. Eventually, under Charlemagne, they were compelled by the hostilities of dangerous, fast-moving mounted opponents such as the Saracens, Vikings and Hungarians to adopt armoured cavalry on a very large scale.

Influential Frankish weapons copied and developed by other tribes included their famous small but lethal throwing axe called the francisca, which Vikings adapted into their larger and popular bearded axe (see Chapter 8). Anglo-Saxons eventually produced the scramasax, which became their ubiquitous dagger, or short sword, from about the mid-7th century, by copying the Frankish seax (see Chapter 6). Finally, the highly efficient Frankish spear, with its large, winged head, was adopted by several tribes, particularly the Vikings.

In the 4th to 7th centuries of the Migration period some Germanic groups in north-west Europe, for example the Angles, Saxons and Franks, did not adopt the military custom of the Visigoths, Ostrogoths and Lombards, who used armoured, mounted cavalry to achieve their numerous victories. 'It would appear that the moors and the heaths of northern Germany and Schleswig and the heaths and marshes of Belgium were less favourable to the growth of cavalry than the steppes of the Ukraine or the plains of the Danube valley.'[1] Perhaps, also, the high cost of acquiring suitable horses and a lack of equestrian and horse-breeding skills were other inhibiting factors. Consequently, Frankish armies were for several centuries mainly infantry ones despite periodic military disasters caused by this principle. Nonetheless, the Franks were a physically robust and brave people, and provided Aetius with excellent infantry at his famous victory over the Huns at the battle of Chalons (451). Eventually the need for cavalry was made very clear to them by their stunning defeat at Casilium (554), inflicted by foes with primarily mounted armies, particularly mounted archers.

'The tactics of their (Franks) foot-soldiery continued for some time to adhere to the old Germanic square, often mistakenly described as a great disorderly mass of

unarmoured infantry fighting in dense formations. Only in the sixth century did they show and signs of Roman influence.'[2]

Frankish Weapons

The Franks were somewhat more diversely equipped with weapons than the Angles or Saxons throughout the Migration period. For centuries their infantry lacked metal helmets or body armour such as mail shirts and greaves, but did possess large, strong, oval shields with a pronounced, central iron boss and edges protected with an iron rim. They carried two spears; one the fearsome barbed angon (see picture overleaf), the other of conventional form. Additionally, they were equipped with a formidable and useful range of ancillary arms such as the throwing axe, short seax sword and long dagger. Also, 'by the middle of the seventh century breastplates had come into general use among the Franks.'[3] They lacked missile weapons such as the bow or slings. Their chieftains were exceptionally well armed with heavily decorated pattern-welded, broad-bladed swords and beautifully embellished short seax swords. 'There were lavishly armed chiefs, as we know from the recently discovered Frankish graves of the sixth century warrior at Morken in the Rhineland.'[4]

'Here the main interest centres on the helmet which was extremely well preserved. It has some features in common with the Sutton Hoo and Vendel types but far more which are quite different: the skull is more conical in form made in several panels set vertically between the spaces of a framework of bands springing from a horizontal band round the brow and joined at the top in a point, where the skull terminates in a little hollow fitting into which a crest, probably a plume, could be stuck.'[5]

Germanic Frankish weapons in our period indicate a significant improvement from those they used several centuries earlier. 'The amount of metal used in them allows two conclusions; firstly, that the iron ore, or metals in general must have been more accessible then; secondly, that some degree of industrial processing must have been achieved.'[6] Certainly, early Frankish arms that I classified were often of acceptable quality.

The Angon

This was a highly efficient combat weapon. It comprised a 3-foot iron shaft tipped with two barbs, set on a wooden shaft. It was 'a dart neither very long nor very short which can be used against an enemy either by grasping it as a pike or hurling it.'[7] It was thus similar to a Roman pilum, from which it was probably copied. This they also hurled accurately at the enemy line, where it generally struck a shield causing the barbs to bend at

A rather fine Frankish sword from Baden Wurttemberg, Germany. Note ovoidal pommel, pronounced iron guard, and fuller. Considerable organic scabbard remains on blade. 7th to 8th C. Acc No: 1908, 8-1,499. (© Copyright the British Museum)

the shank. The angon's weight then dragged down the shield depriving the owner of its protection. The barbs were not easily removed and the iron shaft could not be cut off. At this distracting point a Frank might rush up and kill the opponent with his second spear. Sometimes, the angon would strike an opponent's body and immediately kill him. It might also only wound him but the barbs could rarely be successfully extracted. It will be appreciated that if some four to five thousand angons were simultaneously hurled at an enemy formation their natural reaction would be to raise their shields in a defensive posture. This would frequently allow the barbed angons to penetrate many shields, thus achieving their object. This arm is regularly found at various locations in southern Britain. Seven were actually discovered in the Sutton Hoo grave. This again raises the question whether the Angles and Saxons also made angons or acquired them from the Franks.

Throwing Axes

A particularly popular arm was the francisca throwing axe, which the Franks hurled with great accuracy at short range, often with devastating effect. These are small, ranging in length from the shortest pattern at about 4.75 inches to the longest at about 7.5 inches measured from blade cutting edge to collar back. The cutting blade edges ranged from about 2.25 inches to about 5 inches. All were heavy for their size but well balanced and frequently beautifully crafted. Despite their rather 'dinky' form they were devastating in close-quarter, face-to-face combat.

> 'The francisca, however, was the great weapon of the people from whom it derived its name. It was a single-bladed battle axe with a heavy head composed of a long blade curved on its outer face and deeply hollowed in the interior. It was carefully weighted so that it could be used, rather like an American tomahawk, for hurling at the enemy. The skill with which the Franks

Frankish sword hilt of 6th to 7th C., of high status, decorated with silver inlay and linear decoration. Note stylized zoomorphic animal heads at each pommel-end. Pommel cap secured to the substantial tang-point cover. Blade's upper section covered by wood and leather scabbard remains. (©Copyright the British Museum)

discharged this weapon, just before closing with the hostile line, was extraordinary, and its effectiveness made it their favourite arm.' [8]

I undertook a detailed classification of twenty throwing axes at the British Museum. All the examples classified had been discovered in the areas of the Marne and Rhineland and dated from around the second half of the 5th century to the early 7th century. The various forms (see Figure 1 on page 60) were eventually categorized into six Group Types. The early ones were in rather plain and simple form whereas the later ones were progressively more sophisticated and flamboyant, with sharply upward-curved blade sides and trumpet-shaped heads. Axes found in Britain resembled the earlier forms. Some examples were slightly longer with particularly elegant heads, flowing blades and acutely angled swan-necked shoulders with marked upper blade-tip elevations (see pictures on page 57). Possibly the later designs contributed to the aerodynamic efficiency of the weapon. The Franks constantly practised axe throwing to ensure that the blade cutting edge actually struck a target effectively. The axes were hurled at a maximum range of 30 metres. Within this were three 1.5-metre killing zones, separated by 2.5-metre gaps. The first killing zone was about four metres from the thrower, and within each zone the head of the spinning axe was facing forwards and downwards in its trajectory and therefore caused maximum casualties.

The collar was the most substantial section of the axe, and was designed to retain the

Frankish foederatus wielding angon. His Roman cavalry sword is supported by cross belt bearing a religious rondel and a 'protective' prayer plaque. His helmet is 'a late Roman ridge one' worn by warriors serving with the late Roman army. An example of this helmet was discovered at Burgh Castle, an old Roman-Saxon shore fort near Great Yarmouth. His shield features a substantial umbo (boss), examples of which have been discovered in north Germany. (John Eagle)

Substantial Frankish cone-shaped shield boss (umbo) with silvered copper alloy rivets on flange. The shield on which this was fitted was used both for defence and as an offensive weapon to batter down an opponent and cause facial injuries. From grave 1, Baden Wurttemberg, Germany, 7th to 8th C. (© Copyright the British Museum)

haft (handle), provide the weight necessary to propel the weapon and increase the penetrative force of its cutting edge. Some collar tops were slightly rounded; others were flat. I gave a good deal of consideration to determining how hafts were retained securely within the axe-heads. This was naturally essential to prevent the axe-head being separated from the haft and flying towards the enemy leaving only the handle in a warrior's hand. At first sight the rounded, or rather rounded-rectangular form haft holes appear rather large compared with the overall size of the axe. However, these were deliberately fashioned to accommodate a substantial and robust haft, and thus handhold. The haft hole measurements were about .75in wide and 1in long. Generally, widths were the same at top and bottom but on some the bottom was narrower. This was deliberate to ensure that the haft was very firmly held at the lower end to make its retention more secure. Another way of ensuring firm retention might have been to fit the haft to the collar in the method that wheelwrights use to construct carriage wheels. This principle could have been employed with axes by driving a wooden haft into a heated axe-head collar then cooling this with water causing the iron to contract and bind tightly on the haft. Possibly they may have used a modern method. This involves

Frankish Throwing Axe. Unusual example of short, Type IV early bearded throwing axe, 5th C. to about AD 580. This was the significant forerunner to the later Viking bearded axe. Length about 6in (15.2cm). (© Copyright the British Museum)

cutting a V-shaped slot in the haft top, driving this into the haft hole and then inserting a V-shaped wedge into the V-slot and hammering it down. The tapering wedge ensures constant outward pressure towards the hole sides, thus achieving a firm haft grip. It must have been of consequence that some collar sizes were markedly larger and heavier than those of other types. The reasons for this, perhaps, related to the search for more effective aerodynamic efficiency and the achievement of a longer throwing range.

We now appreciate why the francisca was such a popular Frankish weapon. Its design incorporates knowledge of both fairly advanced iron-processing technology and understanding of the aerodynamic aspects of efficient missile flight paths. Weaponsmiths thus combined embryonic knowledge of technical matters with high artistic skills to produce such a highly elegant and effective short-range missile. Examples of these periodically emerge during archaeological excavations in Britain, indicating a Frankish presence.

The Frankish Seax

In addition to their very-well-known angon and franciscas the Franks employed a variety of single-bladed seaxes from the mid-5th to the early 9th century. I carried out a classification of the large seax collection in the Department of Prehistory and Europe in the British Museum. These dated from the late 5th to the 8th century and are detailed below.

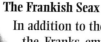

Frankish Throwing Axe. Type I a 'smaller' axe with modest blade similar to the late Roman ones. Note very large haft hole. Possibly AD 500–600. (© Copyright the British Museum)

The title 'seax' is the useful collective word used to embrace various forms of second-line single-edged swords (some broad others narrow), 'choppers' and daggers. Some of these were deliberately dual-purpose, being employed for domestic and farming tasks as well as in battle. Readers are reminded that the well-known Anglo-Saxon scramasax was closely influenced by the Frankish seax (see Chapter 6). The proliferation of these items arose from the desire of Frankish chieftains to equip as many of their followers as possible with a personal edged weapon. The various types were broadly categorized by the typology created by the famous archaeologist K Bohmer who, over many years made a detailed study of Frankish weapons and allied artefacts in the Trier region:

THE SEAX		
Type	*Description*	*Continental Dating*
Class A	Narrow and small seax	Mid-5th to 6th centuries
Class B	Broad seax	7th century
Class C	Long, or lang, seax	8th, and into the 9th century

It is sometimes difficult to differentiate between a narrow, small seax sword and long dagger if both overall lengths are similar. J Giesle's[9] measurement typology scale was thus adopted to ensure accurate categorization of differing but similar arms:

1. Short, narrow seaxes	length: 20 to 25cm
2. Narrow seaxes	width: up to 4cm
3. Broad seaxes	width: from 4cm and greater.
4. Long, or lang, seaxes	length: from 50cm and greater

Broad seaxes are further differentiated as follows:

1. Light ones	length: 28 to 36cm width: 4 to 4.6cm
2. Heavy ones	length: 30 to 42cm width: 4.8 to 5.6cm

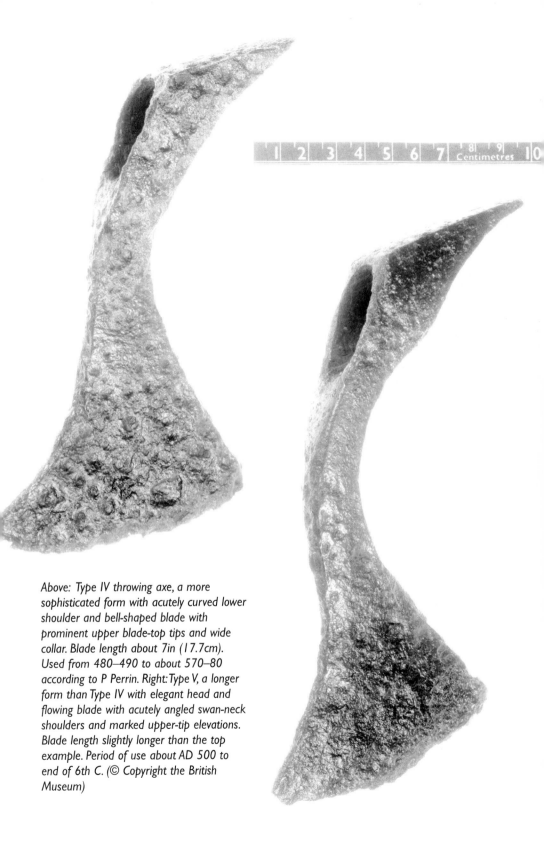

Above: Type IV throwing axe, a more
sophisticated form with acutely curved lower
shoulder and bell-shaped blade with
prominent upper blade-top tips and wide
collar. Blade length about 7in (17.7cm).
Used from 480–490 to about 570–80
according to P Perrin. Right: Type V, a longer
form than Type IV with elegant head and
flowing blade with acutely angled swan-neck
shoulders and marked upper-tip elevations.
Blade length slightly longer than the top
example. Period of use about AD 500 to
end of 6th C. (© Copyright the British
Museum)

Tangs

These were of particular interest, being generally wide, thick, long, and markedly different from and more substantial than those of the Anglo-Saxon seaxes I classified at the Museum of London. The latter were generally smaller and much more delicate. On many Frankish seaxes, especially swords, the tangs were particularly broad, emerging from the blade top close to the blade edges but particularly close to the blade-back side. Several were much longer than those made in Britain providing sufficient space for a two, or even two-and-a-half hand grip. Some long tang examples were probably even longer than those recorded in the classification notes, because several had obviously lost sections of their tang points. These must have enabled exceptionally powerful blows to be delivered. If so, one wonders what the warrior did with his shield in combat when both hands were occupied wielding his sword. Perhaps such grips were actually primarily designed to provide a more effective agricultural chopping tool when warriors were using them on their farms: a good example of an efficient dual-purpose weapon.

The form and design of the junction of the blade and tang was noticeable. On the blade-back top side there is a gentle downward curved (wave) forming the tang top. On the lower cutting-edge side there is a more pronounced feature of a right-angled step from which the lower tang edge continues. Often this 'right-angle' is not crafted in a definitive manner, thus creating a blurred and rather rounded form instead. This indicates poor forging techniques.

Significantly, seax swords were often carried by senior commanders in addition to their broad-bladed ones. These were of higher quality than those used by their soldiers. Ewart Oakeshott states in *The Archaeology of Weapons* that King Childeric, who died in AD 481, possessed a seax. It had a hilt decorated with garnets inset into cloisons of gold and gold foil, and was carried diagonally across his body.

Blades

All blades were single-edged and many heavily corroded. This tends to restrict the classification and dating processes. Their state indicated that most were recovered from damp and wet and, in some cases, acid soil. A significant sample of the collection was examined radiographically. No pattern-welded blade or maker's marks were found. However, several showed clear signs of a 'piled' structure manufacture, indicating that they were made by welding together a number of individual rods of iron, but without the twisting used for true pattern-welding. The term 'piled' structure thus refers to a number of thin, parallel strips of iron welded together by forging. This is thought to avoid the problem of large lumps of slag spoiling the blade and to carburize the iron better. It confirmed a more sophisticated production method than their condition first indicated. Many were heavy in ratio to their lengths, clumsy to handle and sometimes very 'point-heavy'. No complete scabbard was found but their previous existence is confirmed by wood and leather fragments on several blades. Ewart Oakeshott suggests that in the late Roman period seax scabbards were made from 'many longitudinal strips of wood bound at intervals with metal bands'. The broad blade-backs increased the weight of all sword types and generated more powerful slashing blows.

Creation of the Blade Point Section

This area is the most characteristic feature that immediately identifies a seax as being of Frankish manufacture and origin. The section extends from the area at which the blade-back and cutting edge start to bend towards each other. This involves a downward slant of the blade-back joining the upward curve of the blade cutting edge. The overall appearance of the point section is thus of modestly rounded shape (see pictures, right). This is very different from an Anglo-Saxon scramasax, which has a flat cutting-blade edge along the entire blade length.

Eventually the classified weapons revealed the following:

1. Broad-bladed seax swords 21
2. Short, narrow-bladed seax swords 10
3. Daggers (of these three might possibly be short swords) 11
4. Lang seax sword of over 68cm long (Norwegian?) 1
5. Miscellaneous: tang, pommel, chape, knives etc. 9

Total of Frankish weapons artefacts: 52

The above clearly indicates the range and versatility of 'second-line' weapons carried by Frankish warriors, which usefully increased their combat ability. The figure at 3 above confirms the periodic difficulty of differentiating between long daggers and short, narrow-bladed seax swords. The inclusion of the lang seax, which may or not have been of Frankish origin, is an example of the fascinating way unexpected weapon forms suddenly emerge from an arms collection.

Frankish seaxes. Left: broad type with iron binding, presumably from scabbard top, just below the grip. 7th C., found near Mainz, Germany. Centre: broad type with traditional parallel grooves beneath the blade back. 7th to 8th C., from Marne region of France. Right: Narrow-bladed seax-type with remains of scabbard top beneath the tang. 7th to 8th C., from Marne region, France. (© Copyright the British Museum)

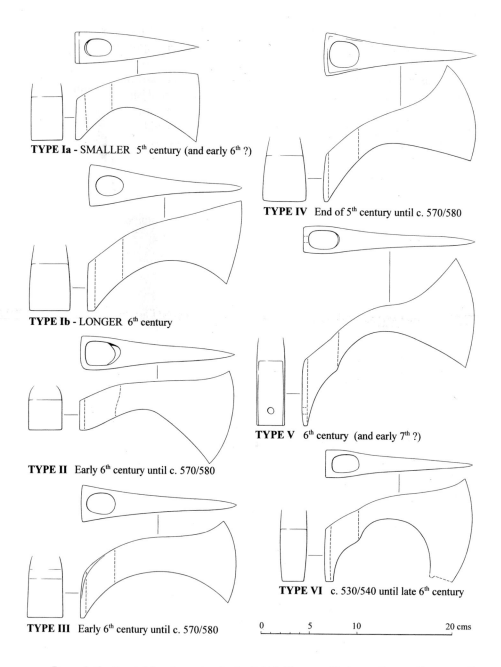

TYPE Ia - SMALLER 5th century (and early 6th ?)

TYPE Ib - LONGER 6th century

TYPE II Early 6th century until c. 570/580

TYPE III Early 6th century until c. 570/580

TYPE IV End of 5th century until c. 570/580

TYPE V 6th century (and early 7th ?)

TYPE VI c. 530/540 until late 6th century

0   5   10   20 cms

Figure 1. Continental francisca types in the British Museum. (By James Farrant, illustrator of The Department of Prehistory and Europe at the British Museum.) (© Copyright the British Museum)

Daggers

Surviving numbers of Frankish daggers indicate that their use was widespread during a period before the Anglo-Saxons were equipped with them. Their appearance is menacing and they would have been effective as an arm of last resort. At the Devizes museum, I classified two Frankish daggers on display. These may indicate a Frankish presence in that area. Both examples have the characteristic Frankish rounded point-section (see picture, right) and should not be confused with Saxon scramasaxes. Their overall lengths are: 22.75 inches and 22.0 inches including a curved iron pommel – a rare surviving feature.

Rare and interesting Frankish dagger (Acc no: DM 64) from Purton cemetery near Devizes, discovered with two knives and a blue glass bead. This dates from about AD 550 to 600, either indicating a Frankish presence in the area or an example of barter trade with the Continent. The curved iron guard is of particular interest, as very few survive on such daggers. Overall length 22in (58.3cm) blade length 14.2in (35.5cm). (Devizes Museum)

Chapter Five

Anglo-Saxon and Viking Period Spears

This chapter kindly contributed by Paul Hill

**The Northmen went in nailed ships
The bloodied survivors of spears on dingy sea,
Seeking Dublin across the deep water.**

From the poem 'The Battle of Brunanburh', Anglo-Saxon Chronicle entry for AD 937.

The spear was a weapon of considerable importance throughout all archaeological periods in Britain. Its most widespread employment was in the period between the end of Roman Britain and the high Middle Ages. Many weapon-heads from the Bronze Age also survive in great numbers, indicating that the spear was not just a ubiquitous weapon, even then, but probably had various functions based on the width and length of its shaft and the design of the weapon-head itself.

The iron spear in the Anglo-Saxon and Viking period was a weapon of the free man, and as such provided him with a status symbol, whatever his rank. A Carolingian law code of the period describes the punishment for a slave found carrying a spear as having the weapon struck across his back until it broke.[1] The different types of spears were employed in various ways in battlefield conditions. The lighter and more fragile ones clearly had an alternative role as a missile, whereas some of the larger weapon-heads had shafts that could only be held in two hands, and had such important fighting characteristics that they would rarely have left their owners' hands in the field of battle.

To give some idea of how a spear ranked among the other weapons in the Dark Age weapon set, it is worth noting that in the wills, or heriots, as they are known (which listed the war gear owed to the lord of the manor on the death of a tenanted thegn, or man of higher rank), from the 10th and 11th centuries, the spear made a consistent and conspicuous appearance. Irrespective of the deceased's rank (right up to an earl), the most requested weapon in terms of numbers was the spear.

The period falls neatly into two groups of surviving evidence, which also mark a change in the nature of the usage of the weapon. From *c*.AD 400 to *c*.700, the spears of the Dark Ages come to us principally from cemetery excavations and stray finds. After the abandonment of the pagan practice of burying men with weapons, the archaeological record becomes less comprehensive and it is difficult to find comparable material with which to date our spear finds. The spears of the period *c*.AD 700–1100 come to us largely through river finds, and some occasional land discoveries which can be described as pagan Viking single burials. The locations of the river finds have prompted some theories as to whether these weapons were lost in battle at important crossing-points (many Anglo-Saxon battles were fought at sites of strategic significance). For example, the wealth of material from the river at Brentford (Old England), which spans most of the Saxon period, cannot have arrived there by accident. There was even a recorded battle there in 1016 between English and Danish forces, the outcome of which was a hard-fought victory for the English, who chased the vanquished Danes into the ford, many of the pursuing English drowning in the process. These later spears are characterized by their greater diversity of form and among them are some of the largest weapon-heads ever known. One such spearhead in the British Museum collection is 80cm long.

Another characteristic of the later period is that the weapons of the Anglo-Saxons, Vikings and, to a certain extent, those of the later Franks, can be seen to cross-fertilize, making it hard to ascertain what is a true Scandinavian, a true English or a true French weapon. The nature of the morphology or form of these weapons is outlined below in Figure 2. Each period is discussed, followed by a section on the method of fighting during the period.

Because of the quantity of material available, the spears of the pagan Saxon period have been exhaustively studied and classified by Michael Swanton.[2] In summary, the key features of these weapons include a distinctive open socket ferrule and weapon lengths which, with the honourable exception of the many long angons known from this period, do not exceed 61cm in length. The open-

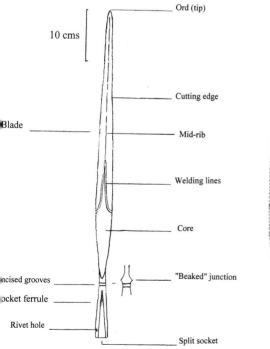

Figure 1. Long thrusting spear. (Paul Hill)

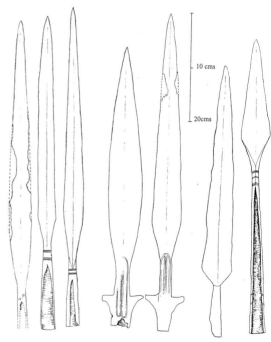

Figure 2. Long thrusting and winged spears from English museums. (Paul Hill)

socket-ferrule method of manufacture during the 5th and 7th centuries is echoed on the Continent in Merovingian France. However, as Swanton pointed out, the method continued in use in England for some centuries after the pagan period, whereas on the Continent and in Scandinavia, the method of constructing the ferrule was closed and welded. The continuation of the open socket ferrule, therefore, is one of the key indicators as to the Englishness of a weapon. For example, in the later period, we might come across two spears of similar length and weight, one with an open socket and one with its socket welded shut. We may reasonably conclude that the former was of English manufacture and the latter of Danish, Norwegian or continental. From the evidence of the male graves in the pagan cemeteries, where spears were laid out in the weapons burial rite, it seems that spear shaft lengths were anything between 6 and 11 feet. This much can be ascertained by the presence in a few graves of the spear-butt ferrule, a small tube of iron retaining the base end of the shaft and indicating the overall length of the weapon when compared with the position of the iron weapon-head in the grave. In Denmark, the famous bog deposits in Nydam, which contain the weapons of warriors from the immediate pre-Migration period, indicate lengths of 8 to 11.5 feet long. So, although there were certainly javelins present in the pagan-period weapon set, it seems that there were also two-handed long thrusting types of spear too.

The rediscovery of trading networks, in particular in the south-east of England in the 8th century at places such as Hamwih (Southampton) and Londenwic, led to the influx of new materials and, in particular, must have introduced English smiths to the new continental weapon types. In fact, Mortimer Wheeler attributed the lengthening of the English weapon-heads in the middle to later Saxon periods to Scandinavian influence, which itself was attributable to the opening of trading contacts with the Rhineland smiths, who were generally thought to be the leaders in European weapon production.[3] The long, narrow leaf-shaped blade of the new era is known to have been exported from the Rhineland to countries as far afield as Finland and Russia.

One of the principal problems that the lengthening of a blade would create is that the junction between the blade and the socket would be the weapon's weakest spot. Many devices appear to have been adopted to solve this problem, not least the introduction of the baluster moulding at this point, but, as some weapons show, even this was not enough to stop a weapon bending on impact. Occasionally, a spearhead might have a reinforcing copper band at the junction, or the amount of metal at the junction would be increased by faceting it. This provided the shank of angled sides, a technique attributed to the Scandinavian smiths.

On many of the spearheads in museum collections around Britain, there is some evidence for damage. Some of this may be put down to post-depositional damage, sustained by corrosive or destructive forces after the weapon was disposed of. However, there are some tell-tale signs of bending and twisting which give clues as to the usage of the weapon. In a recent survey of 200 spearheads in British Museum collections, I discovered that 14 per cent of them have clear evidence of battle damage, the most obvious of which was to the blade of MOL 7769, which was bent into an S-shape as if it had been forced into a shield. Similarly, other blades were bent dramatically at the junction as if they had fallen back on the weight of their shaft after being lodged into a shield.[4]

There are a wide variety of terms in Old English for a spear. In fact, the word 'spear' itself only refers to the iron weapon-head and is one of the least-frequently encountered terms in literature. Other terms were just as specific and indicate that the Saxons and Vikings knew exactly what each type of weapon was used for. The term 'aesc' (pronounced 'ash') was used in literary sources 25 times and refers exclusively to a large, two-handed, long-bladed weapon with an ash shaft, which is very often associated with a wealthy owner, such as a thegn, eldorman or an earl. A warrior from the poem *The Battle of Maldon* (991) is described: 'Byrhtwold spoke up, he raised his shield And brandished his Aesc. He was an old retainer.' Here there is an explicit association with the aesc and the rank of its owner. By far the most common term used in the literary sources, with 78 references, is the word 'gar' (in Old Norse, this word is 'geir' and probably meant the same type of weapon).[5] The word survives today in terms less martial such as garlic (spear-leek) and garfish (a fish with a spear-like nose). In the period in question the weapon is most often described as being retained in the hands of its owners, although it is sometimes described in flight, which indicates that it was of a small enough size to be thrown effectively if required. The aetgar seems to have been a lighter version of the gar and is most frequently described in flight. Lighter still, and clearly a missile weapon, was the daroo, a word from which we get the word 'dart'. The later term 'gafeluc' probably refers to the gar and daroo in flight. Other related terms in this period show some interesting aspects of spear usage. An aescberend, for example, was a spear-bearer. Whether this applied to the owner or indicated that the weapon was of such importance that it had its own carrier is unknown. The tip of the weapon was known as the 'ord'. This term applied to the tip of the wedge-type of battlefield formation, which was an attacking posture, the warriors at the head of which constituted its 'ord'. Some terms indicate the usage of a spear for another purpose: that of hunting. The term 'huntingspere' is known, as is the term 'barspere' (boar spear). The wild boar was an animal of noted ferocity in ancient times and hunting it successfully was something of a rite of passage for a young man. The weapon with which one hunted such an animal was not, as is often surmised, a bow and arrow (for this was the weapon of a poacher): it was, in fact, the boar spear. This type of spear needed to be long in the blade to get through the thick hide of the boar. It may also have been 'winged' near the junction to stop the boar charging up the shaft after being impaled, as some have suggested. The derivation and usage of the strange wings seen on later Saxon and Viking spears is discussed below, but it is reasonable to suppose that the barspere, whatever it looked like, fulfilled another role on the battlefield.

The winged spear (sometimes erroneously called the 'Carolingian' spear after the period in which it seems to proliferate in the archaeological record) appears to have been an item introduced into the Anglo-Saxon weapon set in about the 9th century, probably by the Danish Great Army. The concept of the winged spear, however, may not have been a Danish invention, since it is noticeable in Frankish contexts as early as the 7th century. Even here, the origins are not certain, and it is likely in my view that the weapon first appeared further east, in Byzantium.

The defining features of the winged spear are a set of single wings welded to either side of the socket ferrule. The ferrule is always welded shut, and not cleft or open, which in itself indicates a continental origin for the type. The blade of the weapon is nearly always a broad leaf-shape anything between 400 and 650mm in length. Clearly, these weapons were intended to remain in the hands of their owners and may have had a role as a cavalry lance. Sometimes, manuscript depictions show multiple lines indicating multiple wings at this point, but no spearheads with more than one set of wings have been found. What is most likely is that these lines represent protruding rivets, used to secure the weapon-head to the shaft. The earliest depiction of the winged spearhead is from a relief carving of a mounted warrior dating to *c*.AD 700 from the Magdeburg region, showing the weapon pointing downwards and its head appearing just beyond that of the horse. Throughout the 10th and 11th centuries there were numerous depictions in manuscript illuminations of the weapon type, many of them English in origin, but by far the most explicit example is the illustration from Avranches showing St Michael overcoming Satan, with Satan prostrated, thrusting a typical winged spearhead of the period into his mount.

Before outlining the ways in which spears were used in battle, I shall summarize the types of weapon as we now know them from the literary and artefactual evidence set out in this chapter.

1. Javelins

The light spear, or javelin, was used for throwing, with a range of effectiveness of perhaps up to 25 yards. Defining features included an internal socket diameter of just 9 to 17mm, which indicates a shaft incapable of sustaining lateral damage. A warrior could carry up to three of these in his left hand behind his shield. This is shown by the detail of the Anglo-Saxon shield-wall line, as depicted in the Bayeux Tapestry.

2. Spears

The most common type, mentioned as 'gar' or 'spere' in the sources, it was a fairly light weapon, suitable for being thrown in anger at close quarters, but likely to be retained for hand-to-hand spear play. The most usual internal socket diameter was around 22mm (about an inch), which indicates a shaft thick enough to sustain some lateral damage in use.

3. Long Thrusting Spears

This is a weapon more typical of the later period, but not exclusive to it, and was probably the aesc of the noble retainer. Characteristically, quite a long weapon, it was intended to be retained in the hand. It may have developed from the boar spear and was quite suitable for use by infantry and cavalry alike, but was most probably used in England as an infantry weapon, perhaps in an anti-cavalry role towards the end of our period, when the arrival of Norman cavalry made such employment necessary.

Medley of spears of middle/late Saxon period. From left: 1. From Chiswick Eyot. 2. Open socket without basal moulding. 3. From River Thames. 4. Early 6th to late 7th C., flattened blade, angular shoulders and split socket. 5. Plain spearhead with slight beak at blade base. 6 larger spear with horizontal and slightly 'beaked' moulding at blade base. 7. Open socket with grooved junction. 8. Large spearhead, much decayed but retaining pronounced 'beak' at blade base. 9. Large spearhead with traces of horizontal moulding and pronounced 'beak' at blade base. (Museum of London)

4. Heavy Spears

These merit a special mention, being similar in form to the categories of both spears and long-thrusting spears, but all of them having an unusual weight, commonly over 500g. They were used in the same manner as long thrusting spears and were not thrown in battle.

5. Winged Spears

A Scandinavian addition to the English weapon set, they were derived from spears which originated in Francia, and probably ultimately derived from ones made in Byzantium. A weapon designed for effective defence and probably employed in the shield wall by noble retainers in the same way as the long thrusting spear.

How were the battles fought between armies of the period armed predominantly with these types of spears? There is much compelling literature on the subject, not least this passage from the poem *The Battle of Maldon*:

> Likewise they were encouraged by Aethelgar's son Godric,
> onward to the struggle. Often he sent a spear, A slaughter-shaft,
> spinning into the Vikings...

Leaving aside such emotive words, more practical help comes from the activities of some re-enactors of the period who have tried to identify the weapons'

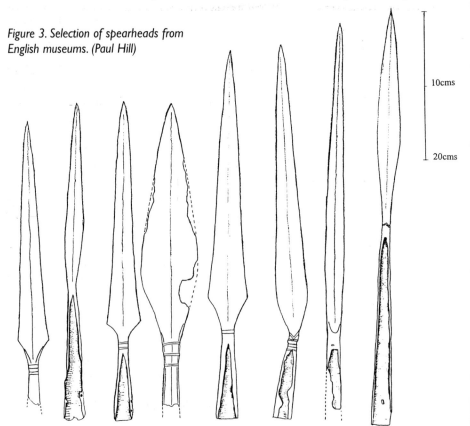

Figure 3. Selection of spearheads from English museums. (Paul Hill)

10cms

20cms

characteristics in flight, when retained in the hand and when used in close-order shield-wall work. Many spears, it seems, can be wielded one-handed, provided that they are light enough. These are used in conjunction with the shield and are often wielded overarm. The Bayeux Tapestry shows spears employed in this way, even though this was in an age when the larger long thrusting spears were readily available. The great advantage of spears is their ability to keep the enemy at distance. Foes with swords and axes, whilst devastating at point of contact, are less effective if they are kept at spear's length. The effect of interlocking shields and long thrusting spears in the front rank must have been impressive indeed, provided that the solidity of the formation could be retained. The shield wall when advancing was often likened to a steam-roller and must have echoed the power and density of the ancient phalanxes of Greece and Macedon. Shield walls need not be tightly packed (as they were at Hastings, for example). When closely ordered, the shields would have been interlocked rim-to-boss. When a shield wall fought in a more open style, perhaps over difficult terrain, the shields would have been at best rim-to-rim.[6]

Re-enactment has revealed some surprising facts about spear play. Kim Siddorn, one of Britain's leading re-enactors, has pointed out that each spearman in the shield wall has three enemies, the one in front of him, the one on his shielded side, and the one on the right, the most dangerous, who can see through the gap on the spearman's unshielded right-hand side. Once an enemy sees this gap and exploits it, a shield wall can collapse in a very short space of time, particularly if swift-moving, lightly armoured foes can penetrate the gap. More often than not, we are told, you do not see the person who 'kills' you.

Some observations have been brought to light by those who wield replicas of the winged spear. Some arguments that have been put forward are based around the idea that the wings are there to stop the weapon travelling through an opponent. They are too deeply positioned for this, however. Many of the blades on this type of spear are so long that they would have gone right through an enemy before the wings would take effect. Those who use these replicas have commented that even a small wing can be used very effectively to parry or trap an opponent's spear or sword, allowing a shield-wall companion to capitalize on his misfortune. Some skilled spearmen would have been able to hook their wings around opponents' shields and open them up in this way. In effect, the winged spear was a spear with a built-in hilt, or guard.

The devastating effect of a javelin thrown at close quarters should not be underestimated. It need not be razor-sharp to be effective. In fact, few, if any, show signs of having been sharpened, although it is entirely possible that this evidence, being only superficial evidence, will have eroded away. Re-enactors have shown that a javelin hurled at a pig carcass, even when protected by a chain-mail outer layer, would still penetrate to an impressive 13cm depth.[7] It is easy to see how bodies of light infantry armed with such weapons could have softened up a shield wall before it came into contact with its main enemy. Still more effective would be the added velocity of a javelin hurled from horseback. The Anglo-Saxons and Vikings, however, are not known to have employed such a tactical unit, as both armies were composed largely of infantry.

The change in the nature of the spear over the period AD 400–1100 reflects a

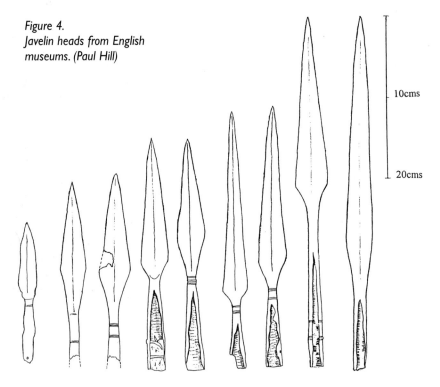

Figure 4.
Javelin heads from English
museums. (Paul Hill)

10cms

20cms

general change in the style of English warfare. In the early Migration period, the Germanic communities who buried their weapons in their pagan cemeteries generally fought in an aggressive and attacking manner. Their ferocity was well known in the late Roman world, and as tactical fighting units their numbers were small. In fact, estimates from the Danish bog deposits suggest that the average Germanic war band of the period may have numbered fewer than 200 men. This did not detract from their effectiveness in the field, however. It is exactly these kinds of war band who fought in Britain in the first generation against the Picts and other enemies of the Romano-British establishment, using their naval strength as well as their land-fighting capabilities to protect the eastern seaboard from raiders. The buried weapon sets of these warriors sometimes do not seem to reflect the true fighting practices of the time, owing more, perhaps, to ritual than to actuality. Some warriors, for example, were buried with only a single shield. However, taken as a whole, the weapon set does indeed present an image of an attacking posture for these forces. The shields of the early period were all small, with a protruding boss, indicating use as a buckler in conjunction with a sword. The swords themselves, as outlined in Chapter 4, were remarkably effective slashing weapons. The spears in this era, whilst there are a few long thrusting types, as one would surely expect, show a tendency towards javelins and spears, indicating a greater degree of tactical mobility for the infantry units on the battlefield.

As the English settlements slowly became the English kingdoms, during the 6th and 7th centuries, so the armies recruited from them changed in character. If manuscript depictions from this later period as a whole are to be believed, the shields got much bigger and the long thrusting spear and its variants, including the winged spear, came more and more into play as the style of tactical fighting became increasingly defensive. It was in the later period that the shield wall (or 'bord wall') came into its own. The Anglo-Saxon army was no longer a small, closely related unit of warriors from just a handful of kin groups, but was being recruited from wide areas of territories owing military service to each king.

In 865 the Danish Great Army came to England and within a generation of their stay had utterly transformed the political and military situation. Of the seven Anglo-Saxon kingdoms from the golden age of the heptarchy, only one survived – the resourceful kingdom of Wessex. Only Alfred the Great's ingenuity in the face of disaster saved England from becoming Danish. The Great Army had split up into northern and southern branches and this in the end was probably part of its undoing, as Guthrum the Dane led a surprise winter attack at Chippenham and from there sortied into Wessex, where Alfred brought him to a decisive battle at Edgington in 878.

The key to understanding how warfare was waged in this period of the Great Army is hidden within the following statement in the Anglo-Saxon Chronicle entry for AD 871.

> During this year, nine general engagements were fought against
> the Danish army in the kingdom south of the Thames, besides
> the expeditions which the king's brother, Alfred and single
> ealdormen and king's thegns rode on, which were not counted.

It is clear that the noblemen of England were expected to ride after their foes, indicating that they were effectively mounted infantry. The nine general engagements mentioned would have been those which the main army, recruited from the whole of the kingdom, fought against the main Danish host. The Anglo-Saxon army was indeed an infantry army, despite many of its more elite troops owning and using horses.

The age of the spear in England did not exactly die in 1066. The arrival of the Normans, with their couched-lance fighting style, did change things a little, however, and over the next 100 years, fighting styles changed to those of armoured, mounted knights, with the spear as only one of a great variety of weapons available to the feudal warrior.

Chapter Six

Germanic Ancillary Weapons

This chapter examines additional essential accoutrements carried by Germanic warriors during the Migration period. Anglo-Saxon and Viking spears and javelins are described in Chapter 5 and their formidable axes in Chapter 8. Here we will look at the shield and the Anglo-Saxon scramasax dagger. Some shields were specifically designed for use as both a defensive aid and a weapon to batter down opponents in a manner similar to that employed by Roman soldiers. Many later shields were designed purely for defence. Both types can be regarded as indispensable accessories that provided significant protection and a useful second-line offensive weapon. As I have said, in contemporary society:

> 'Possession of arms was also important, perhaps even more so than for the nobles. The bearing of arms, like the length of hair was in German society a symbol of legal freedom. The freeman's arms, as the Carolingian capitularies reveal, were the shield and the spear: arma id est scutum et lanceam.'[1]

SHIELDS

In this year king Athelstan, lord of warriors,
Ring-giver of men, with his brother prince Edmund,
Won undying glory with the edges of swords,
In warfare around Brunanburh.
With their hammered blades, the sons of Edward
Clove the shield-wall and hacked the linden bucklers.

The Anglo-Saxon Chronicle AD 973

Migration warriors attempted to protect themselves with a small shield or buckler carried on the left arm. This afforded some shelter from missiles, spear-thrusts and sword-slashes. It was their primary defensive aid, as very few men at that time wore the expensive and rare iron helmets or coats of mail. These were not to become fairly common until about the late 9th century. Early round shields did not, however, compare in size or effectiveness with the large, straight-sided and convex ones used by Roman legionaries, which provided considerable body protection and efficient means of belabouring opponents.

Shield Designs
These normally comprised a round, flat disc or orb, made from alder, willow or linden (lime) wood, and were carried on the left arm. Early-pattern shields were just under half

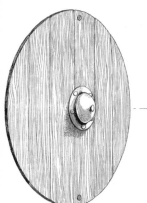

Reconstruction of an Anglo-Saxon shield. (Museum of London)

an inch thick (5 to 9mm), but later-pattern examples may have been thicker. I examined a small section of shield wood which surrounded a surviving rivet, and it was exactly half an inch thick.

Evidence of shield diameters 'suggests they were generally smallish and circular, and could be divided into three "ranges": the smallest 13.25 to 16.5in (34–42cm), the middle range 17.75–26.0in (45–66cm), and the largest from 27.5 to 36.25in (70–92cm).'[2] Presumably, the younger, smaller men carried the little shields and the stronger men the more substantial ones. They were constructed from small planks of hardwood and on a very small number the edge was reinforced with a metal rim. Through the centre of the shield a hole was cut which was then covered by a forged iron boss and its flange riveted to the wood board on the outer side. This afforded security to the left hand and knuckles on the inner side. Also on the inner side was fitted a shield-handle crossbar made from wood-covered metal. This was occasionally bound with leather or cloth. Härke suggests a detailed classification of the lengths of shield grips (handles) into the following groups: short 4 to 6in (11–16cm), medium 8 to 10in (20–25cm) and long 12 to 16in (30–40cm). Most of those I worked on were of the small size. 'Existing remains of English shields point to a construction of boards laid side by side, not overlying each other like plywood although this has been a popular assumption for many years.'[3] It seems from Scandinavian evidence that shields could be made from at least three planks and 'the evidence suggests that as many as nine might have been used; they were probably fixed together with glue and ledges, the planks rebated along their edges to form a flush surface, strengthened towards the middle by the long grip (on the inner side) and riveted across three or more.'[4]

A few of the later shields were slightly curved, and this was achieved simply by a steaming and bending process. Research and discussion has been devoted to clarifying the extent of convex shield carriage sometimes depicted in Saxon manuals, and in one frame of the Bayeaux Tapestry, but evidence on this point is not conclusive because so few curved shields have survived.

By the second half of the 11th century the very-well-armed English regular army, comprising the elite housecarls and thegns maintained by the King and nobility, carried the large, sophisticated and robust kite-shaped shield. These are frequently depicted in the Bayeux Tapestry. They were employed by both the English and the Norman French when mounted because they provided good protection to the left side of a horseman's body. When fighting dismounted the housecarls also used them, sometimes lightly dug in, to enable the axe to be wielded in both hands. This shield is studied further in Chapter 11. The earlier and more simply fashioned round shield continued to be used by most members of the Great Fyrd and these are also illustrated in the Bayeaux Tapestry.

The weapons of the Scottish, including their shields (targes), were sometimes leather-covered. It is possible that some Migration period ones were also covered with bull's hide on the front. A shield from Sutton Hoo incorporated leather covering. This could have been fitted damp in one piece, which, when dry, would shrink and thus strengthen the whole artefact. Shields were fitted with a strap: perhaps they were carried on the warrior's back like those of the Highlanders in the 17th and 18th centuries.

Shield Bosses

Shield bosses are the most distinctive and fascinating characteristic of the shield. Fortunately, because they were constructed from iron, in a most skilful manner, many have survived. They were produced in varying forms, and their very long, rather complicated evolution makes them particularly fascinating to study. As I mentioned, these protected the warrior's left hand, head and upper body when fighting in either a defensive or offensive manner. The shield boss typology diagram highlights the various boss forms. Shields in the early pagan Saxon period were generally smaller than those of later centuries, comprising a flat round board probably of plank construction with a central aperture over which an iron boss and grip arrangement was riveted. Shield bosses of the earliest type (Dickinson and Härke 1992, Group 4) have a tall, straight-sided cone rising to a sharp apex rod. These early bosses indicate that shields may have been used offensively, the whole shield being employed as a buckler, the hand protected or enclosed by what literally amounted to a spiked iron fist.

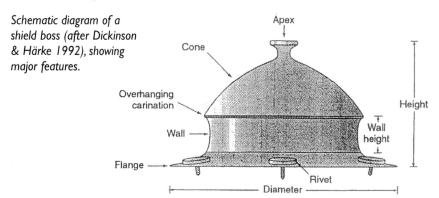

Schematic diagram of a shield boss (after Dickinson & Härke 1992), showing major features.

'Boss forms of the later 5th and 6th centuries are not as high-sided as the earlier ones (at least not until the later 6th century when the conspicuous sugar-loaf variety makes an appearance) and many have domed or convex cones indicating a possible switch to the defensive in fighting styles. These frequently have a small disc at the tip of the apex.'[5]

The shield boss typology diagram on page 75 shows that:

'Group 1 comprises generally low bosses of the 5th to 6th centuries with overhanging carinations, concave walls and either straight (Group 1.1) or convex cones (Group 1.2). Group 2, which run throughout the 6th century, shares elements with both Groups 1 and 3, and are characterised by straight-sided walls, and are more common in East Anglia and the West Midlands than in areas south of the Thames. Group 3, a popular group, are narrower and taller than their forebears, and have a date ranging through the whole of the 6th century and probably into the 7th century. They retain the overhanging carination but have a convex cone with straight-sided walls and are frequently found with five flange rivets. Group 3 bosses are thought to be the influence for the later low curved varieties of which Park Lane cemetery

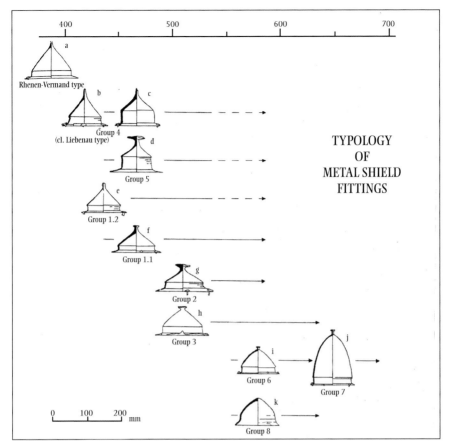

Schematic summary of typological and chronological sequence of shield boss groups in Early Anglo-Saxon England: a. Richborough inner ditch (after Hawkes and Dunning 1961, fig. 5d); b. Long Wittenham I, grave 138; c. Abingdon I, grave B33; d. Sutton Courtenay II, unassociated; e. Berinsfield grave 69; f. Cassington II, grave 2; g. Abingdon I, grave B39; h. Bidford-on-Avon grave 69; i. Taplow (after Evison 1963, fig. 12d); j. Lowbury Hill (after Evison 1963, fig. 26g); k. Bergh Apton grave 19 (after Green and Rogerson 1978, fig. 76Fi; n.b. this boss has been chosen for illustration because it represents a distinctive subtype within an otherwise rather miscellaneous group) (approx. 1:8). (Dickinson & Härke, 1992)

had one example (ON 52, grave 74; figures 19 and 60) with no apex stud probably of the later 6th to early 7th century date.'[6]

'A boss and its accompanying flange was secured to the wooden board with very broad-headed bronze rivets fitted through holes in the flange. Sometimes, these rivets, which may be of slightly different size, are silvered, a sure sign of status. The author examined one boss, accompanied with five rivets only one of which was silvered. Doubtless its owner was delighted to possess at least a single silvered one. During classification of weapons from the Park Lane cemetery in Croydon, the author detected on a Group 3 shield boss (ON 140 from grave 147) two clear damage marks on the cone-two straight cuts, 59mm and 42mm long – which were probably made by a sword since a spear would have produced a perforation. A most exciting

Ten shield bosses covering period 5th to 6th C., recovered from early pagan Saxon cemetery at Park Lane, Croydon by Wessex Archaeology during 1999 and 2000. (Photograph by Elaine A Wakefield reproduced by courtesy of Ms J I McKinley, senior archaeologist of Wessex Archaeology, Salisbury)

moment. One cut was deeper than the other. These marks instantly create an exciting vision of close-quarter combat. This damage supports the notion that Group 3 bosses were used in a more defensive manner than the earliest types of shields, their curved sides helping to deflect the blows. Such marks are rare in bosses of this period. The Park Lane damage marks constitute vertical slashing impact to the cone and fit in with the contention that Group 3 bosses reflect a change in usage over time.'[7]

Shields in Combat

It will be appreciated that, until the early 11th century when, perhaps, more warriors of the Select and Great Fyrd possessed some improved forms of protective clothing, the shield was their only means of defence. Because these could only protect a proportion of the body at any one moment it must have been necessary in battle to move the shield promptly to block sudden blows from unexpected directions. On occasions, combat survival must also, to an important degree, have depended on athletic ducking and weaving to avoid some potentially fatal blows. Defence against swords was the most difficult to achieve because these could be wielded with a medley of different devastating strokes at close range. A warrior had thus to be prepared to counter sweeping blows against his lower legs, direct thrusts with tapering swords towards his midriff, horizontal slashes towards his face, and powerful over-shoulder slashes aimed at his head and shoulders. The sword and axe were probably the most effective weapons in causing serious damage to shields, which were frequently destroyed in battle. Shields were often painted to conceal the line of the grain. If this was horizontal, the shield could be easily split by an axe or sword.

Shields were an important and generally effective means of creating a strong 'shield-wall' defence line, which was the basic principle of Anglo-Saxon battle tactics. Evidently, this involved the men standing closely together holding their shields at the correct height and angle with their spears jutted forward between the shields to present and intimidating array of blades. The long spears of the men in the second rank would further contribute to the hedgehog-like appearance of the front line. Generally, if this line held, the battle would eventually be won; if it was broken the scattered survivors could more easily be slain in a rout, which would further encourage their surviving comrades to flee.

SCRAMASAXES

Origins and Classification

This weapon and/or tool was initially introduced by the Franks during the 5th century, and was much developed thereafter (see Chapter 4).

> 'The term "scramasax" is used to denote Frankish and other single-edged knives, or cutlasses because of an oft quoted passage in Gregory of Tours, who, in his history of the Franks speaks of "boys with strong knives (cultris validis) which they commonly call 'scramasaxes' (scramasaxos)".'[8]

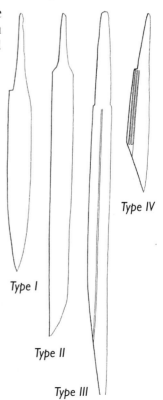

I own an Anglo-Saxon 3-inch knife, in miniature scramasax form, which was probably used for kitchen tasks. The word 'scramasax' is, like the Frankish word 'seax', actually a collective name for a wide assortment of edged instruments. It thus refers to a very numerous assortment of Anglo-Saxon edged tools, weapons and implements made in different sizes for diverse purposes. The scramasax, in more modest form, was eventually inherited from the Franks by the Anglo-Saxons in the late 7th century and used as both a weapon and farm implement.

Thereafter, the adoption of the long scramasax and many other forms of dagger or long knives in varying lengths gradually became widespread in Britain and remained so for about four centuries. Scramasaxes were very often carried by warriors in addition to the spear and also by those armed with a sword.

Type IV

Type I

Type II

Type III

Classification of scramasax development.
I: Frankish type, late 6th–8th C.
II: Norwegian type, 7th and 8th C.
III: Hurbuck type, 8th–10th C.
IV: Honey Lane type 10th–11th C.
(Museum of London catalogue)

Saxon warrior of early 10th C. with light spear and scramasax dagger of the Honey Lane type. (John Eagle)

The scramasax collection in the Museum of London contains a profusion of differing types, but they may be categorized as follows: considerable weapons of more than about 2.5 feet long were classified as swords; ones of less than 7 inches as household utensils, and those in between as daggers or long daggers. It is stressed that these measurements relate to Anglo-Saxon weapons and not to Frankish ones. According to D M Wilson in his book *The Anglo-Saxons*, 'the scramasax do not appear in England until the latter part of the pagan period.' 'The scramasax proper – proper at least in archaeology, may best be described as a sort of clumsy carving knife, to which on the average its size likewise approximates.'[9] These arms are often noticeable for their sturdiness, blade thickness, and rather heavy weight. They are also often clumsy to handle, and often have a tang that seems too short in ratio to the blade. Anglo-Saxon examples differ from Frankish ones in that the blades are often lighter and shorter, the tangs much shorter, and generally the whole weapon is smaller and sometimes more neatly fashioned. The most significant difference between Frankish and Anglo-Saxon versions is the form of the point section. Frankish ones are invariably slightly rounded owing to the junction of a downward curve of the blade-back and an upward curve of the blade-edge to form the point. Anglo-Saxon ones generally have a shorter point section due to the downward-curved blade-back joining the flat blade edge (see picture on page 80). The rounded form of Frankish short swords and dagger-blade points is a useful feature with which to identify them (see pictures of Frankish ones in Chapter 4). Ellis Davidson, in *The Sword in Anglo-Saxon England*, says of the scramasax and handseax: 'There are a number of very varied examples of these in English museums, and some have Anglo-Saxon or runic inscriptions.'

Seax and Scramasax Typology

The seax was introduced by the Franks in the late 5th century in Gaul and Germany, and then accepted later in Scandinavia. It was eventually adopted in Britain in the late 7th or early 8th century. During this period its form, appearance and size periodically changed. However, scramasaxes of all categories were always single-bladed. The typology diagram on page 77 shows that there are four main basic types, but to these must be added numerous other forms resulting from regional and transitional changes. Indeed, there are probably more non-standard forms than standard scramasaxes. This point can initially cause some confusion when identifying such arms.

Type I: Frankish

Seaxes of this type, which includes swords and daggers (handseaxes), were examined at length in Chapter 4 and above. In particular, we noted their rounded point sections, created by the junction of the downward-curving blade back joining the upward-curving cutting edge. Early patterns were first introduced about the mid-5th century in Merovingian Gaul. Some, from the late 6th or early 7th century, have been discovered in England. Devizes museum has two long daggers, one with a curved iron pommel, which I dated at about AD 600.

Scramasax swords and daggers. From left to right: Type III short sword or long dagger; Type I Frankish sword; Type I Frankish short dagger, Type I and III transitional short sword; Type I Frankish narrow seax; Type III sword; type IV robust scramasax tool and weapon; classic Type IV (note parallel lines at blade top); transitional type; a mix of Types I and III. (Museum of London)

Type II: Norwegian

These were produced during the 7th and 8th centuries. The most distinctive feature of this single-edged sword is the straight, flat-topped blade, which is sometimes rather convex at the blade top centre. This type (very similar to the Frankish lang seax) was common in Norway, where narrow, single-edged swords were very popular. The majority of their seaxes were thus swords, some of them being very long and narrow-bladed.

Type III: Hurbuck

This type was made and used in Britain from the 8th to the 10th century. The name relates to a 9th century hoard found in the location of that place name in England. Note the break of the blade top, which then stretches down at a gradual angle to form a very long blade point. There are deep parallel lines etched below the blade back top on both blade sides. Such areas were sometimes inscribed with copper, brass or white metal. The majority of this category must have been swords.

Type IV: Honey Lane

This type was produced in Britain during the 10th and very early 11th centuries. Their distinctive shape was created by a very sharp obtuse downward angle of the blade top joining the flat blade cutting edge to form a sharp tip. An example of these was found in the Honey Lane market, hence its type name. This example was dated by means of coins found with the arm that were dated between AD 978 and 1016. These daggers were invariably decorated towards the blade tops with a series of parallel lines. This area was also heavily decorated with intricate engraving. Many daggers of this type have tangs that extend parallel with the blade edges. They were carried in angular leather sheaths. 'Between this extreme form and the Hurbuck type are many transitional and cross bred examples; but the scarcity of evidence as to date prevents us from arranging these in a reliable, evolutionary sequence.'[10]

Throughout the Anglo-Saxon era scramasaxes of various forms were in use simultaneously owing to the slow evolution of weapons and the natural conservatism of those who wished to retain an old but still serviceable arm to which they were accustomed.

Construction

Scramasaxes were usually made of ordinary iron. Some better-quality ones were produced from 'steely iron', and a few of high quality were created by pattern-welding. This involved the laborious and skilful task of twisting together iron bands of differing qualities, heating them, and then hammering and constantly turning the mass into a blade, and finally, hardening the cutting edge. It is also very possible that some were made from previously produced pattern-welded bars. Lang and Ager compared the construction of a seax with that of a sword:

> 'Most of the Anglo-Saxon seaxes required less time to construct as they were more simply made. Seaxes only have one cutting edge, are usually shorter in length, require less metal and probably had a handle which was simpler than the more complex sword hilt. Strengthening was given by the heavy back edge. As long as this was sound the weapon could be used.'[11]

This weapon or tool ranged in length from about 3 to 29 inches, and the longer ones were actually short swords. The majority of blades were heavy in relation to their length. The blade tops were particularly thick and robust, which contributed to the clumsy feel of the weapons.

Thus they were often point-heavy. However, these handling shortcomings did create one combat advantage, as we shall see later. Tangs (handles) were naturally created during the weapon's construction. Some of these were definitely rather too short, and narrow in relation to the blade weight and length, being shorter and narrower than many Frankish ones. Their characteristics were thus rather similar to many early Migration period swords.

I have not seen a guard on any Anglo-Saxon scramasax, which may indicate that they were not used. If some were used they would have been constructed of wood or horn, which is unlikely to survive. Iron guards were periodically seen on Frankish seaxes. Hilts would have been formed from semi-circular cheek plates of wood or horn, possibly glued or bound to the tang sides, which may have incorporated a small, rounded, wooden pommel. To provide a more comfortable handhold the tang grip would then be covered with leather and tightly bound. Pommels were varied in shape and were often small and unobtrusive. It is possible, however, that sometimes the whole tang was enclosed in a carved-wood hilt that incorporated a low guard at the front, a hand grip, and a rather larger, rounded pommel. This feature was noted on some Frankish seaxes. On some fine examples, a more elaborate pommel of iron would be incorporated in 'cocked-hat' shape during the weapon-forging process. This could sometimes be finely and heavily decorated with gold and silver (see picture on page 83). When the dagger or short sword was sheathed, the pommel still remained in view, so this visible area was an important potential place to exhibit status. Consequently, owners spent as much as they

The Sittingbourne seax. An extremely high-status Honey Lane type of the early 10th C. Decorated on both blade sides. Ellis Davidson states 'it has panels of mosaic inlay and champleve work resembling that on some of the more elaborate sword hilts.' Example decorated with copper, bronze, silver and niello depicting foliage and beasts in Trewhiddle style. Facing side states: 'Gebereht' and other letters meaning 'Gebereht owns me' and the other side: 'Biorthelm made me.' Neg no: AS 20. 95. (© Copyright the British Museum)

could afford on pommel decoration. A scramasax was usually carried either horizontally across the stomach, 'or was suspended from the waist belt by means of a series of small bronze loops.'[12]

Combat

The weapons of most Anglo-Saxon warriors comprised the shield, spear, javelin and scramasax. Unlike the javelin, the spear was not hurled at opponents, because such action would naturally deprive the warrior of his main arm. In contemporary battles, men did not deploy in the regimented, highly disciplined interlocked ranks so splendidly demonstrated by the Roman army. Instead, they fought in a close-quarters mêlée that was a chaotic, confusing and dangerous situation where numerous individual man-to-man conflicts took place simultaneously. The shield provided vital protection while the spear was vigorously directed with overarm or upward thrusts at any weakly guarded parts of an adversary's body. It was a period of stimulating, adrenalin-pumping struggle that demanded great courage. Survival also depended on physical fitness and the ability to duck and weave to avoid an opponent's lunge and recover to an advantageous position from which a decisive stroke could be delivered. If the spear was broken, accidentally dropped, or became too firmly embedded in an antagonist's shield or body the warrior was momentarily defenceless. He then resorted to employing his scramasax if there was no discarded spear immediately available that could be safely snatched up and used. Long-bladed scramasaxes could be wielded in over-the-shoulder or sideways arm-slashing strokes, the heavy blade inflicting significant wounds. Its long, pointed blade, reinforced by the very heavy blade back, also made it a good thrusting weapon guaranteed to penetrate contemporary protective clothing and cause deep, fatal wounds.

We have now examined the Anglo-Saxon scramasax and noted how it was a popular second-line arm for some four centuries. Its retention for such a long period is proof of its effectiveness and dependability. Apart from successful combat use it was doubtless employed in a multitude of tasks which have occupied soldiers since ancient times. In this, it was rather similar to the 19th century sword bayonet, which was employed for woodcutting, carpentry, game butchery and cooking. As a supplement to the spear and shield it was a versatile, useful and practical weapon.

Winchester seax or scramasax dagger with fairly complex pommel and thin oval guard, both silver plated, indicating a status arm. Either Saxon or, possibly, Frankish of 9th C. (© Copyright the British Museum)

Chapter Seven

Viking Warfare and Weapons

'From the Fury of the Northmen, O Lord, deliver us'

The main purpose of this chapter is to examine Viking swords, but first, we look briefly at their magnificent longships, the essential means by which they accomplished piratical expeditions, ambitious trading ventures, conquests and eventually, widespread colonization. Without these ships it is unlikely that Scandinavians would have escaped the confines of their generally inhospitable homelands and become, for centuries, the scourge of Europe. 'Obviously it was the Viking ship that made the Scandinavian break-out possible.'[1] The Viking race mainly comprised Danes from Denmark, Norsemen from Norway, large communities in Sweden, and the Finns. Additionally, their numbers 'were increased by many other adventurous folk who joined them on their expeditions and became part of their society.'[2] As Scandinavians, they were related to the Anglo-Saxons and shared with them common attributes. They possessed, however, additional outstanding characteristics that gave them justifiable claim to be classed among the most dominant races in world history. Their tremendous energy and determination in all they did made them brave and capable soldiers, hardy and skilful mariners, conscientious farmers, ambitious traders, and highly artistic and ingenious craftsmen. 'The so-called Viking Age began around AD 800 and lasted for nearly three centuries.'[3] During this period they utilized their navies to undertake ruthless plundering and conquering expeditions. Evidence of the casualties of their raids, the widespread pillage and devastation they inflicted, and the high numbers of victims they supplied to foreign slave markets, confirms that pagan Vikings, and indeed some later Christianized ones too, really were ruthless and cruel in their dealings with others. The western European populace, particularly in unprotected coastal areas and those adjacent to major rivers, were understandably obsessed by the threat Vikings posed to their moderately peaceful existence. Eventually Viking raiding progressed to conquest and finally, colonization, of parts of England, Scotland, Ireland, The Faeroes, the Isle of Man, France, Iceland and Russia.

The Ships

The greatest Viking love was for the sea and their magnificent and graceful longships, the designs and construction of which always reflected the latest advanced technology. Such beautiful vessels were the aquatic symbol of their civilization, and were created with the care and artistry with which a more static society would build its public monuments. The oak hulls were strong but flexible, and ships had pine masts from which hung rectangular double-layered sails made of tightly woven cloth. The wool used came from an ancient breed of sheep that had a double fleece comprising soft wool next to the skin and an outer layer of stronger, waterproof wool. Yarns were made from each layer and then woven together to produce a tightly woven, water-resistant and hard-wearing material. The main advantage of the final cloth was its considerable elasticity, which enabled the sail to be used higher into the wind.[4] The efficient rudder was sited at the steerboard quarter. The high prow was frequently finished with carvings in the form of dragon- or animal-heads. According to Magnus Magnusson in his book Vikings, the surviving Gokstand vessel provides us with much valuable information. 'This was about 23.5 metres long, 5.2 metres broad and 1.95 metres deep from gunwhale amidships to the bottom of the keel. It weighed over 7 tons yet even when fully laden with another 10 tons of equipment and freight it drew only about 1 metre of water.'[5] The ships were fast owing to a combination of an aerodynamic design and clinker-built hull construction. This caused some air to be drawn under the hull, thus minimising drag and turbulence and causing the prow to rise through the waves in a planing motion. This enabled at least some ships, when the wind was behind them, to travel at up to 16 knots.[6] Their shallow draught enabled them to be beached, making harbours unnecessary, and sail up major rivers to reach inland cities such as London, York and Paris. With the best warships in both northern and southern waters, the Vikings were able to embark with impunity on their expeditions. Nautical supremacy and the advantage of surprise generally enabled their fleets to attack when and where they wished. These benefits were enhanced by their ability to land almost anywhere due to their ships' shallow draughts. If, on berthing, an invasion force was confronted by a superior host, a rapid withdrawal by sea would be ordered, and landing made elsewhere. Vikings realized that land mobility could be increased with horses, and to achieve this, would steal mounts in the vicinity of their landing places. They were thus masters of the art of achieving tactical surprise and employing flexible tactics.

Viking Weapons

'A man should never move an inch from his weapons when out in the fields, for he never knows when he will need his spear.'

Havamal, a poem of Viking Age origin

During the Viking era, fighting was mostly undertaken at close quarters in a desperate adrenalin-pumping mêlée of hacking and thrusting, ducking and evasion. For this, the Vikings, some of whom were full-time professional warriors, were generally very well armed. They carried wooden-shafted spears with a range of leaf-shaped iron heads and shaft sizes, which could be used for a variety of tasks. This type was the most common and popular Viking arm (see Chapter 5). Taller men tended to carry longer spears, and the weapon's main purpose in battle was to keep opposing warriors who wielded dangerous swords at a distance. Spears were also used for hunting, thus providing a dual-purpose implement. Shorter and lighter, carefully balanced ones with narrow tapering heads were javelins designed to be thrown at opponents from a distance. Heavier ones were employed for thrusting at close quarters. The one shared characteristic was that they nearly always had a closed, hollow tubular socket into which the wooden shaft was inserted and nailed. This feature was unlike the open-cleft ones of the Anglo-Saxons. A later type, presumably copied from the Franks, was of winged form with two pronounced, metal projecting wings designed either to deflect an opponent's spearhead or to prevent the blade penetrating too deeply into an adversary's body and delaying its withdrawal. Such instances could momentarily make a Viking vulnerable.

Another favourite weapon was the axe. These had been copied from the devastating and famous small throwing axes (franciscas) of the Franks used during the 5th to 7th centuries. The Vikings also copied the Frankish small 'bearded' axe and developed this into their own much larger 'bearded' axe, which became very popular. Many warriors carried them during the 8th and 9th centuries and employed them with great ferocity. Some warriors used bows and arrows for hunting and fighting. Surviving examples are rare; however, 'a massive longbow recently excavated at Hedeby would have been a weapon of considerable power.'[7] Some arrowheads in leaf or angular shape have been found including an especially penetrative type designed to pierce armoured helmets and ring-mail. Only a few Vikings initially possessed iron helmets because they were very expensive. Occasionally these were fitted with nasal guards and rather strange spectacle-type eye-guards. Mail shirts were likewise little worn in the early Viking period owing to their cost. However, by the 9th century both helmets and mail shirts were much more common. Mail was often worn over a padded jacket, which provided effective additional defence against spear thrusts.

Wooden shields of various sizes were the standard means of defence. They were round, made generally of lime-wood with a large, circular projecting iron boss on the outer side designed to protect the hand grasping the shield on the reverse side. The substantial boss could also be employed as an offensive arm by pushing and battering down an opponent. Sometimes edges were strengthened with an iron strip and usually the shields (see Chapter 6) were painted with crude symbols to aid recognition in battle and disguise the grain line.

Viking Swords

We examine now in some detail the most significant symbol of Viking power – the single-handed, double-edged sword. To a marked degree these were influenced by Migration period swords made by Dark Age smiths. As always in weapon evolution there was a gradual transitional phase between the use of an old weapon and its replacement by a new one. This was the case between the swords of the later Migration period and those of the early Viking era.

Mortimer Wheeler asserted that 'The heavy cutting sword which was adapted and developed by the Vikings of the 8th century was essentially a weapon of mid European origin.'[8] From the late 8th century their swords gradually developed in a more individual Scandinavian form although still influenced by regular imports of very high-quality Frankish sword blades from the Rhineland. These sword blades were described as: 'very high-quality sword blades imported from the Rhineland

and used for some of the finest Viking swords.'[9] Vikings often fitted their own hilts to imported blades.

The overall sword lengths varied up to about 36 inches, and blade lengths from about 30 to 33 inches. Blade widths just below the guard were about 2.2 inches. However, in the 8th century, blades gradually became wider and heavier, gently tapering to a very slightly rounded point. Their form and design is arresting and rather beautiful. Ewart Oakeshott noticed that 'The Vikings themselves had much more down-to-earth decorations upon their swords – inlay or platings of silver, bronze, tin, copper and brass – which must have been far more hard-wearing.'[10] Occasionally the iron pommel-ends were fashioned with stylized zoomorphic animals. The tell-tale chevron blade patterns indicating the pattern-welding process can often still be seen. Blades were usually fullered with wide, shallow troughs designed to lighten the arm, increase blade flexibility and concentrate the central sword mass towards the blade cutting edges. Some fullers were of different form, comprising several narrow but deeper grooves. Oakeshott observed that 'The further development, roughly coinciding with the beginning of the Viking period, was the substitution of a narrow metal cross-piece (guard) for that of wood and bone.'[11] This more efficient and permanent feature was welded to the blade's upper section providing reliable hand protection. Guards were generally in boat-shape form with rounded ends. They were still rather short and projected only a little beyond the blade edges. Nonetheless, they were much more dependable than the bone or 'sandwich'-type guards of the Migration period, and doubtless more reassuring to the warriors holding them.

Pattern-Welding

The best Viking sword blades were generally still pattern-welded. This is not surprising as it was a logical continuation of a very successful technique undertaken by highly competent Germanic swordsmiths since before the 3rd century AD. Pattern-welding is a complex, scientific metallurgical subject about which much has been written. Here a more succinct explanation will be sufficient. In the Dark Ages it was an efficient means of removing slag from poor-quality iron ores. The working principle adopted was 'to twist together iron rods, and then to weld these rods into a blade-shaped blank which is then further treated to produce the finished piece. Its derivation is from an earlier process whereby alternating qualities or iron sheet or strip were forged-welded together then beaten into final shape.'[12] Aside from the swords, pattern-welding has been found on spearheads and seaxes. It is a time-consuming and labour-intensive technique, but the blades produced by this method are robust and strikingly beautiful (due to the curly serpent blade pattern). Oakeshott noted that 'They are also quite exceptionally supple, which would make them difficult to break in normal use; springy blades were prized for this reason.'[13] Later a wide blade cutting edge of carbonized steel was added to the blade core and welded in position, then filed down carefully to form the cutting edges and point. Many Viking sword blades were engraved with pagan inscriptions.

During the 7th century this manufacturing procedure reached its highest level but during the mid-8th century the practice declined on account of larger supplies of very-good-quality ore and more sophisticated and efficient furnaces. These enabled larger volumes of high-quality steel to be made more rapidly. The lengthy pattern-welding process, which required outstandingly capable swordsmiths, was thus less needed, though it was retained for high-quality sword inscriptions and used much later in Britain.

Increased Sword Demand

During the mid-Viking Age , demand for swords increased considerably owing to more young men wishing to take part in overseas piratical adventures. This conveniently coincided with the major growth of Scandinavian iron-mining and smelting industries in the 7th and 8th centuries. Consequently, sword production increased tremendously leading to a profusion of differing types and patterns of varying quality. To save sword production time, one can assume that many blades were made of the more easily produced 'steely iron' – a process in which at least some carbon was absorbed by the iron mass during forging, and fairly evenly distributed throughout the blade. From then on, fewer Viking swords, with the exception of those used by kings and war band leaders, were so boldly and lavishly decorated with gold, silver and niello as they had been during the Migration period. Instead, and doubtless in part due to the greatly increased work pressure within the sword industry, many were enhanced in a much simpler manner with brass, copper or silver inlay. Indeed some swords had no decoration at all. But, as Oakeshott commented, 'even the plainest iron sword was a supremely functional object of devastating efficiency. It was designed as a single-handed weapon and had blades about three feet long. To be ready for use, a sword was carried in a scabbard often finely ornamented with metal mounts slung from a belt of a baldric across the shoulder.'[14] Viking sword lengths and weights now varied far more than those of the Migration period, when the long, clumsy, broad-bladed, parallel-sided swords usually dictated carriage by a warrior of large physique.

Sword Balance

Most Germanic Migration period swords had either long, broad, parallel-sided blades or slightly tapering ones. Both types were point-heavy and rather clumsy, which meant their point of balance was low on the blade. The former was designed to deliver powerful and devastating over-shoulder blows, whereas the tapering ones, which were easier to manipulate, could also be wielded from above the shoulder or to deliver straight thrusts to the body and face.

The early Viking swords (like those of the Migration period) also had the characteristic of low points of balance. It is thus of great interest and significance that in about the middle of the 9th century there was a significant change in sword blade design, making more of them easier to handle in combat. Ellis Davidson, in *The Sword in Anglo-Saxon England*, states: 'The swords of this time are found to be better balanced, with the centre of gravity no longer half way down the blade but near to the hilt, so that they must have been more efficient weapons for both cutting and thrusting.' It also made them easier and less fatiguing to handle, particularly when lifting the weapon above the shoulder, while being more suitable for delivering effective straight thrusts at an opponent's stomach and horizontal slashing strokes at his face. The new, more versatile blades had another important advantage in that a warrior could very rapidly adopt an effective defensive posture to guard his face and frontal body.

Most Vikings were very effective warriors owing to their regular involvement in both physical and weapon-handling training and their zeal for fighting. These activities were also pursued conscientiously as a means of developing team spirit. Their pagan religion also contributed to their combat-effectiveness because particularly brave battlefield deeds leading to death were the essential means of reaching Valhalla – the Vikings' heaven. No wonder they were so feared.

The Typologies of Viking Swords

Two typologies relating to Viking sword hilts used from about AD 750 to 1100 provide useful information on many employed by the Vikings in this era. They particularly assist in sword type recognition and dating. By the late Migration period, sword hilts had rather larger, stronger but more simply constructed pommel forms than the previous smaller, complex and 'fussy' ones of the mid-Migration period, which frequently incorporated pommels and guards in 'sandwich' sections, often retained with long, gilded bronze-alloy pins. However, we should remember that a few of the very early hilts were notable for their very sophisticated designs and extremely costly gold-plate decoration. The most popular early Viking pommel forms were in triangular shape, as reflected in Petersen's typology at A, B and C. 'Cocked-hat' types, used during the Viking period, were periodically used in larger and more compelling forms. Oakeshott noted that 'Two basic factors are common to all of them – the continuing use of the combined upper guard and pommel, and the extreme development of the latter.'[15]

Petersen's Typology

The first, and still the most influential typology was that devised by the Norwegian Dr Jan Petersen (see Figure 1 overleaf). This study was largely concerned with all sword examples discovered in Norway but also includes swords in several European and continental museums. His detailed and clear analysis is based almost solely on hilt forms and some scabbard-mount ornamentations. It also differentiates between arms with single- and double-edged blades. His work was published in Norwegian in the book *Der Norske Vikingsverd* in 1919. Thereafter it became the standard and authoritative work on the subject, and remains an invaluable guide today. However, it is now in need of updating and this is evidently underway. Please do not be deterred from studying the work because it is not written in English. It is profusely and beautifully illustrated and clearly depicts a large variety of hilts. The evolution and typology of these is consequently fairly easy to follow.

Petersen's method of creating his typology was firstly to select the 26 seemingly dominant sword-hilt types and assemble them in chronological sequence. These were each denoted by a letter from A to Z with the letter J omitted from the sequence. At the end he inserted the final letter 'AE'. The ligatured AE is counted as a single letter at the end of the alphabet in Scandinavian languages. The arms were thus broken down into very narrowly different categories. The 26 hilt examples, with 'A' representing the earliest, were then examined carefully and described in detail. Another twenty swords, which resembled some of the main ones closely or very closely, were called 'saertyper'. He considered these sub-types to have hierarchical similarities. They were not listed in alphabetical order but do appear in numerical order when appropriate. For instance, sword B had one saertype, denoted as B1. Not all main swords had a saertype but apart from B with 1, some others were allocated as follows: G had 6, K had 5, M had 4 and Y had 4. The prodigious number of different sword types in the Viking era clearly demonstrates Scandinavian, continental and Anglo-Saxon determination to fashion pommels in a distinctive pattern.

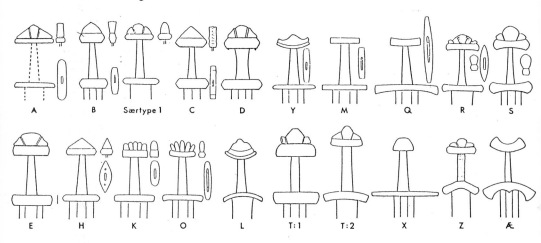

Figure 1: This diagram depicts the most significant hilts from Petersen's typology . Other, lesser-known hilts are not included. (From J Graham Campbell (1980) 'Viking Artefacts', a select catalogue (BM Press, London). (© Copyright the British Museum)

We can also deduce that hilts of varying patterns were being carried simultaneously. Many were taken abroad by warriors, and some have been recovered in the UK, Ireland and in Europe. The swords mentioned by Petersen were all used by Vikings, but not every one was produced in Norway. For instance, his Type L, and its other associated English swords with their curved guards and magnificent Trewhiddle decoration were made in England, whilst many others were fashioned on the Continent, particularly in the Rhineland. Some of the Type K versions came from there. Probably the best Viking sword incorporated a high-quality Rhineland blade. There does not yet seem to be general agreement about the extent to which Vikings employed swords made outside Scandinavia. It is established that sword blades were imported by Scandinavia and that complete swords, made outside Scandinavia, were taken back there by the Vikings. It is hoped that the research on Petersen's typology will clarify this point. The exciting prospect remains, however, that some of the Norwegian-made swords not yet discovered outside that country may suddenly emerge in England. Presumably, also, the superabundance of differing weapon types confirms the very high demand for swords in Norway. They were now prolific and no longer regarded as a rare and elite arm as in the Migration period. It is not surprising that far more Viking swords than those of the Migration period survive today, not only because more were manufactured but because there was naturally slightly less time for them to be corroded.

'From the Viking age more material has survived, and Norwegian sword hilts have been divided into 26 main classes in *Der Norsk Vikingsverd*; although modifications are necessary in considering swords outside Norway, this remains the most valuable classification of hilts in this period.'[16]

Petersen's typology has always been of great value to researchers of the Viking period, and its use by students of such weapons today is still strongly recommended.

Wheeler's Typology

Mortimer Wheeler, in his catalogue 'London and the Vikings', written in 1927, explains his much-abbreviated Viking sword hilt typology (see Figure 2 on page 93). It was partially based on Petersen's typology, but he considered this to be unnecessarily elaborate for the examination of Viking swords and the allied material discovered in England. I think this view is rather limited for such an absorbing and potentially large subject. Mortimer Wheeler's typology comprised seven hilt forms listed by Roman numerals. Later, Ewart Oakeshott helpfully added two extra categories numbered VIII and IX. He rightly thought these were necessary to explain the hilt types and forms during the transition period between the late Viking era and the early 12th century. During this time the Viking sword characteristics of short guards, lobed pommels and traditional blades merged with subsequent early 'knightly' swords with longer, narrower guards and heavier wheel or beehive-shaped pommels. Consequently, some of these 'intermediate' period examples can initially be rather difficult to identify. A good example of this style-matching process was a sword from the R T Gwynn collection sold by Christie's in April 2001. This has a Viking double-edged 10th century blade (repaired) with wide fullers and a later, wider 11th century guard with a flat, hemispherical pommel in semi-circular form. Oakeshott certainly simplified the recognition of the sword types of the transitional period, but it is still possible to find rare examples whose correct identification will require much thought. Since Wheeler published his typology in 1927 there have

An early 11th C. 'intermediate' sword with double-edged, fullered, and possibly old Viking blade with replacement long cross guard of square section and substantial beehive-shaped pommel. Overall length: 38in (96.5cm). Oakeshott Type X. (Crown Copyright, Board of the Trustees of the Armouries)

been many major changes in weapon research. Conservation techniques have improved immeasurably resulting in more vital facts emerging. Further, the use of high-quality X-rays now reveals valuable data about blade-manufacturing methods and the presence of inscriptions. Such information was not available in the 1930s.

It is, perhaps, an appropriate time to consider a little updating of Wheeler's typology, which is rather short in relation to the large number of differing hilt examples. Also, it currently only applies to swords discovered in the UK up to 1927. In this book, however, changes are restricted to altering the typology order and amending the group descriptions of his Types V and VI. Firstly, his typology I diagram is confusingly placed, because he states that this hilt form extends back continuously to the prehistoric Iron Age. This is correct but its inclusion as the first category of Viking sword-hilt types ignores the fact that the traditional Viking Age extended from about AD 800 to 1100. Therefore this sometimes causes researchers, including me, to initially think examples of this form are much earlier than they really are. Similar hilts of predominantly triangular form were used during the 8th century, including the start of the Viking era. Therefore Wheeler's Type I data must be changed.

Secondly, his typology details of Type V are incorrect – his 'Wallingford' example (Petersen's L) was not a Viking sword but a high-quality English one like the Gilling West sword, with distinctive, high-coned pommel (see Chapter 10 for a description of the Gilling sword). Ellis Davidson, in *The Sword in Anglo-Saxon England*, comments: 'Wheeler calls the acutely curved hilt the Wallingford type after the well known hilt in the Ashmolean which was thought until recently to have been found at Wallingford, but is now known to have come from Abingdon.' The latter is actually later than the L Type and is examined in detail in Chapter 9. As there is a sword in the Reading museum called the

Left: Double-edged Viking sword with wide fullers. Petersen Type M and amended Wheeler Type III. Straight guard inlaid with close-set vertical brass threads. Flat-topped pommel matching guard design. Note upper and lower brass ferrules at ends of tang providing extra support to the grip, now missing. Blade length: 28.25in (73.2cm). (Christie, Manson and Wood)

Right: Viking sword with tea cosy or beehive-formed pommel. Petersen Type X and Wheeler's Type VII. Both boat-shaped guard and pommel are finely decorated in the design of quatrefoils with silver lines. The blade, which may actually be earlier than the sword furniture, is pattern-welded. (Christie, Manson and Wood)

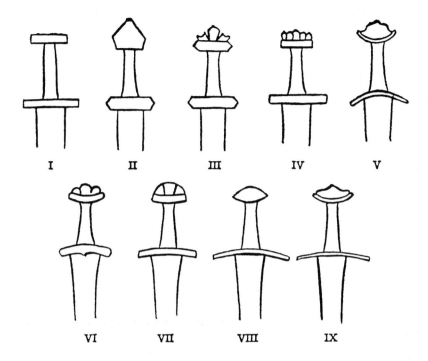

I   II   III   IV   V

VI   VII   VIII   IX

Figure 2. Wheeler's Viking sword hilt typology with Oakeshott's two additions.

Figure 3. Revised typology.

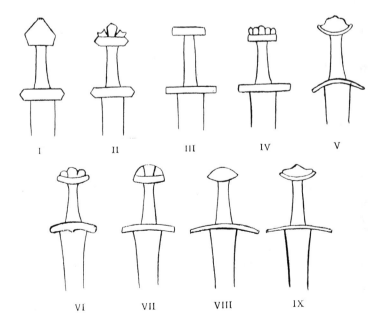

I   II   III   IV   V

VI   VII   VIII   IX

'Wallingford Bridge sword', the further confusion can be appreciated.

In the amended Wheeler typology diagram his Type I hilt has been moved to Type III, where the hilt form equates to the 8th and 9th centuries. His Type I hilt is thus replaced by his Type II showing one of the popular triangular pommels. His original Type II is substituted with his original Type III, which had a three- or five-lobed pommel. Comments on each of the nine types, including numbers VIII and IX introduced by Ewart Oakeshott, are given below. Many of Wheeler's original comments are retained.

REVISED TYPOLOGIES

Type I
The first pommel is of the triangular, simplified cocked-hat type relating to Petersen's Type B. The popularity of the triangular shape is confirmed by Petersen's additional types at A and C. These are clearly Norwegian, employed very early in the 8th century, and are more often found on one-edged blades than on two-edged blades such as the lang seax. Over 330 examples have been found in Norway, a smaller though considerable number in Sweden, but very few in Denmark. These occur in Britain along the routes of early Viking expeditions to the Orkneys, the Western Isles and Dublin. Few have been found in England.

Type II
This form has a three-lobed (occasionally five-lobed) pommel, often zoomorphic. The central one is always the largest and the guard is straight. This type is normally found in north-western Europe dating from the 9th and early 10th centuries but is rare in the British Isles. They occur on the island of Eigg, in Scotland and Dublin. The type's main development was in north-western Germany and Scandinavia under the influence of the zoomorphic pommel studs which were a characteristic of this region in the 5th and 6th centuries. This important pagan religious feature is retained on many examples dating to the end of the 10th century. It may be observed, however, that the cocked hat of the 7th century, almost invariably flanked by two large rivet heads, had independently established a tripartite division of the pommel and so probably contributed to the development of this complex series. On some cocked-hat examples the flanking rivet-heads survive more as ornamental bosses, or as vestigial reminders of the practical importance during the Dark Ages of the gilded pins that helped retain the sandwich-form pommel and lower guard. It is of great interest that during the transition to the three-lobed pommel the zoomorphic tradition was maintained by forming the flanking lobes in exaggerated knob-like animal snouts.

Type III
This simple, rather crude type consists of two straight, matching parallel iron crosspieces employed in the 8th and 9th centuries. There are two examples of this in the Museum of London. Its use continued in Norway as late as the 10th century. This form is a Petersen Type M.

Right: 9th C. sword of Petersen Type K and Wheeler Type IV with double-edged pattern-welded blade with wide, shallow fullers. Narrow cross guard with four short silvered pins, vestigial reminders of the previous manner in which bronze alloy gilded pins retained pommel sections of Migration period guards. Pommel and guard decorated with very close-set silvered lines. Blade length: 30.75in (77.9cm). (Christie, Manson and Wood)

Type IV

This has a five-lobed pommel with the central one slightly higher than the others. It is of the Petersen Type K. This form was probably Frankish made in the Rhineland, where the Vikings acquired many of their best sword blades and doubtless hilt furniture as well. There is an excellent example of one in the Wallace Collection (No. A456) with steel blade and signed guard dated about AD 900–950. A detailed description of this very fine arm is included in Chapter 10.

Type V

The high domed pommel of this hilt was designed and manufactured by English swordsmiths and decoration experts during the approximate period AD 860–890. The production process included pattern-welded blades. This is Petersen Type L. Three examples of the category were found at Gilling West, Wensley and Bedale in Yorkshire, the latter two in Viking graves. The central lobe is particularly large, prominent and high-peaked, taking advantage from its use of the entire tang length. The upper guard is now markedly upwardly curved. Conversely, the rounded lower

A large and significant hilt of a 10th C. Viking sword from the River Thames near Temple Church. The large three-lobed pommel has ornately fashioned linear rope style and silvered decoration. The two outer lobes are distinctly zoomorphic, perhaps representing fierce dogs. The tang grip of silver wire, which has survived, would originally have covered the horn or wood tang-plates, which have disintegrated. Petersen Type S, Wheeler Type VI; late 9th C. (© Copyright the British Museum)

guard is most acutely downward-curved. The decoration in the Trewhiddle style, which is particularly English, is of silver plates chased with diamond patterns set in groups of four. However, neither upper nor lower guard is decorated. The design and decoration is somewhat over-large and exaggerated but in stunning and unmistakable form. No wonder the Vikings wished to acquire this weapon (see Chapter 10 for a full description of the Gilling West sword). Type V is not a modification of a Viking one but originated and was manufactured by the English in England. Their distribution was in England and Norway. The Type V could perhaps be confused with the later, famous Abingdon hilt in the Ashmolean Museum. At first sight there are some similarities. However, the Abingdon hilt is different in the following respects: the lower guard is flat-faced, not rounded, to facilitate its decoration; the cross guard is less acutely curved, being crescent-shaped; the upper pommel guard section is now also decorated in the Trewhiddle style.

Type VI (10th–11th century)

Mortimer Wheeler chose as his example of this category a fine sword discovered in two pieces in rather strange and lucky circumstances from the Thames at Wandsworth (see picture on page 116). This is kept at the Museum of London. It has a three-lobed pommel with distinctive central lobe flanked by two smaller ones reflecting earlier zoomorphic types of animal snouts. The small guard is of flat and triangular form, very similar to the guard discovered in Exeter inscribed 'LEOFRIC ME FEC (IT)' which is now in the British Museum (see picture on page 117). The substantial blade was later discovered to be inscribed with 'INGELRII'. Similar hilt patterns were also fitted to other discovered swords, such as the Brentford, Shifford and Temple swords. The triangular guard form was also used on Scandinavian swords such as those found at Telemark, Buskerud and Oppland. It is suggested that this sword type was predominantly influenced by English swordsmiths, who initially designed the hilt form with its triangular-shaped guard and fitted them with imported blades, on which the inscriptions eventually became more stylized and incomprehensible.

Type VII

This has the late Viking period semi-circular pommel sometimes referred to as 'mushroom' or 'beehive'. Occasionally the pommel is plain, but generally they are marked with shallow grooves or beaded lines that divide the plain surface into three segments. These are obviously vestigial reminders similar to but later than those of the Type II characteristics, with their pronounced lobes, the end ones of which have zoomorphic connotations. The segments of this type each represent one lobe. This sword type is linked to Scandinavia in the 10th century. Examples have been found in York, which was captured by the Danes in AD 867. Others have been found in the Thames, the River Nene near Horsey and the River Lea at Edmonton (now in the Wisbech Museum).

We now examine the additional group Types VIII and IX, which were wisely added to Wheeler's typology by Ewart Oakeshott to cover the transitional period between

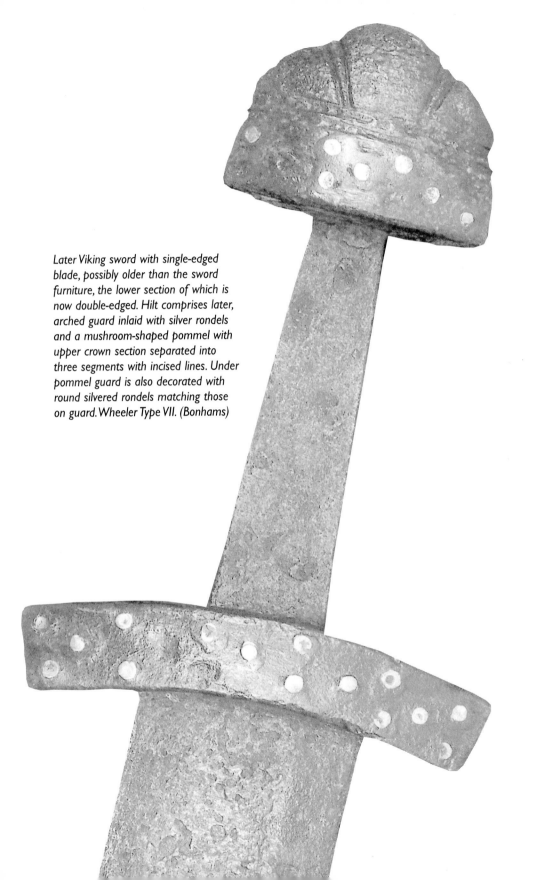

Later Viking sword with single-edged blade, possibly older than the sword furniture, the lower section of which is now double-edged. Hilt comprises later, arched guard inlaid with silver rondels and a mushroom-shaped pommel with upper crown section separated into three segments with incised lines. Under pommel guard is also decorated with round silvered rondels matching those on guard. Wheeler Type VII. (Bonhams)

the late Viking Age swords with their short guards and lobed pommels and the early 'knightly' swords of the late medieval period with crude circular pommels, longer guards and eventually longer blades.

Type VIII
The pommel seems to be a simplification of that on Wheeler's Type VI. According to Oakeshott: 'The divisions between the upper and lower parts have vanished as well as the lobes thus leaving a form just like a Brazil nut.' It will be noted that the guards are longer and narrower and somewhat curved, thus very different from the Viking examples. Such weapons have been found in Viking graves in Norway.

Type IX
Type IX is similar to Type VIII except for the pommel reverting to a small but pronounced cocked-hat form. Such swords have also been recovered from Viking graves of about AD 1100 in Norway. Ewart Oakeshott believes this to be a bye-form of Type VIII with a similar hilt but with a pommel that retains two features: the upper pommel section is an exaggerated cocked-hat shape; the lower a distinctive, uplifted form of the lower section of the earliest 'Brazil nut'.

Chapter Eight

Viking and Anglo-Saxon Axes

This chapter examines the general evolution of axes in Britain from Roman times onwards. Most were designed predominantly for civilian functions, and it is sometimes rather difficult to differentiate between those intended for civilian use and ones for military tasks. We will concentrate on Viking axes, as they were the race that regularly used this weapon during their campaigns in England. Military axes were thus predominantly Viking, and not Anglo-Saxon, arms. We consider, in particular, the large, broad (Type VI) battleaxe – the terror weapon of the 11th century. Detailed specifications are included for the benefit of military modellers, war gamers and re-enactment society members.

The Axe in Action

On 25 September 1066 at Stamford Bridge near York, King Harald Hardrada of Norway, with his ally, the traitor Earl Tostig, brother of King Harold of England, were caught napping by the unexpected arrival of King Harold's army. Hardrada's force was split, with one-third of his host detached at Riccall protecting his fleet. Those under his immediate command were scattered and unprepared on both banks of the unfordable River Derwent. English troops attacked at once overwhelming scattered Viking detachments on the west bank. The English were then confronted with the problem of crossing the deep river and establishing a bridge-head on the far bank. The only means was a wooden bridge defended by a small number of Vikings. Several Englishmen rushed across but were cut down. One Viking wielded in both hands a huge battleaxe with an 11-inch blade mounted on its long haft, which furnished a 7-foot clearance sweep. Another English group approached, more cautiously, to deal with the fearsome axeman. Like their comrades, they were driven back in a welter of severed arms, heads and broken weapons. Then a wily housecarl floated down-river in a malting tub (no boats were available) and while the warrior was distracted, thrust his spear up through a gap between the bridge planks and killed him. English troops stumbled forward over their slain or fearfully wounded fellows and reached the other bank.

EVOLUTION

The vast majority of axes in Roman times were employed on household, farming or artisanal tasks, and thus were tools, and not weapons. Once a pattern was deemed suitable for a particular job, changes were rarely effected and the same axe form could continue for years. This was logical in carpentry work, where tasks remained

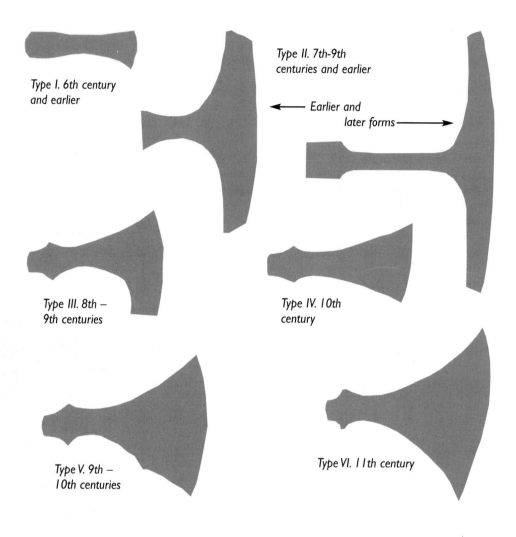

Type I. 6th century
and earlier

Type II. 7th-9th
centuries and earlier

← Earlier and
      later forms →

Type III. 8th –
9th centuries

Type IV. 10th
century

Type V. 9th –
10th centuries

Type VI. 11th century

Figure 1. Types of axes used in the Viking period. (London and the Vikings catalogue,
Museum of London)

similar for generations. Conversely, military axes were subject to regular
modifications to improve their military effectiveness and maintain parity with
those of their enemies.

The Museum of London has an excellent and varied collection of both civilian
and military axes, including large Type VI axe-heads like those used by housecarls
at the battles of Stamford Bridge and Hastings, and many splendid Viking
examples recovered from the Thames. Mortimer Wheeler summarized this subject
in his typology, published in the Museum of London's *London and the Vikings*
catalogue.

Figure 1 relating to the types of axes used by the Romans and Vikings depicts:

Type I. A small Roman civilian axe that was used for woodcutting and was retained in similar form for many centuries. Precise dating of such pieces is difficult, as examples of this Roman one were made over a four-hundred-year period. A significant Roman axe was the bipennis, a two-bladed one used for tree felling.

Type II. A T-type shaped Frankish axe that was also a civilian tool used for carpentry tasks. There are many forms of these, from very large to small, which were used for centuries.

Type III. The Viking 'bearded axe', of which there were several patterns. The Type III is illustrated in Figure 1, and Figure 2 shows the typology of this special axe form. The graceful axes at A to D are Frankish. The first three were small but still fairly substantial throwing axes used by Frankish infantry with significant effect during the 5th to 8th centuries. This arm was one of their favourite weapons and was sometimes specially weighted, like a Red Indians' tomahawk, which they threw with astonishing accuracy at an enemy's head, body or shield. The bearded axes D and E in Figure 2 (our Type III in Figure 1) were used in significant numbers by the Vikings during the 8th and 9th century and these have been discovered in England and the Scottish Islands.

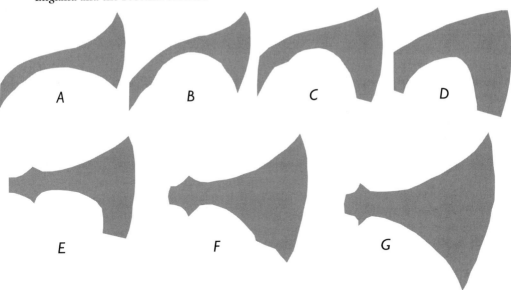

Figure 2. Evolution and devolution of the bearded axe: E, 6th to 11th C; A, C and D, from Herpes, Charente; B from the Marne region. (London and the Vikings catalogue, Museum of London)

Type IV. In the second half of the 9th century, the Type III axe (as shown in Figure 1 and E in Figure 2) gradually merged into the Type IV, with its forward-facing blade. Mortimer Wheeler says of this: 'The more or less symmetrical axe with moderately expanded blade, and spur-projections both above and below the socket, an old type which increased in popularity at the end of the 9th century and

Early 11th C. Viking axe Type V fitted to brass socket ornamented with lines and dots. Discovered at north end of site of the old London Bridge; one of a group of battleaxes and spears found that may represent a battle in the bridge's vicinity during renewed Danish attacks on England in late 10th / early 11th centuries. (Museum of London)

became the characteristic form in the 10th century.' This was possibly used for both military and civilian tasks.

Type V. Figure 1 shows the development of an axe incorporating characteristics of both Types III and IV, with the 'beard' replaced by a wider, more symmetrical blade. Note how the lower blade section, replacing the 'beard', is created by two differing angles, and that the forward section is less pronounced.

THE MANUFACTURE AND SPECIFICATIONS OF THE TYPE VI AXE

'By 1000 AD the standard Viking two-handed battle axe in use in the West seems to have had a graceful flared head (unbroken by any beard or angle) from a narrow neck at the shaft to a broad, curved blade, equally balanced above and below the point of junction with the shaft.'[1] Detailed study and handling of the fascinating, large and diverse axe-head collection of Type VI axes (Anglo-Scandinavian battleaxe) at the Museum of London revealed much interesting information about their construction. All the examples studied were discovered in the Thames, or local London areas.

Most of the large Type VI axe-heads showed sophisticated design and manufacturing techniques in their production. The heads combine strength, balance, powerful and pronounced cutting edges, and unexpected lightness. On the best-made examples it seems that construction was done in two stages: the haft socket and blade back was made in one

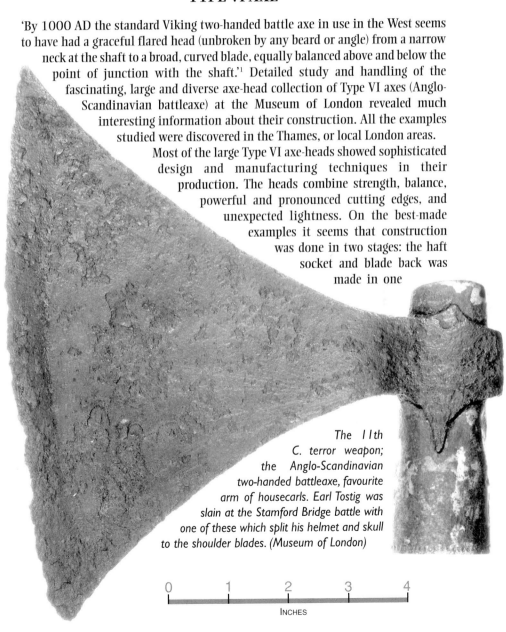

The 11th C. terror weapon; the Anglo-Scandinavian two-handed battleaxe, favourite arm of housecarls. Earl Tostig was slain at the Stamford Bridge battle with one of these which split his helmet and skull to the shoulder blades. (Museum of London)

0   1   2   3   4

INCHES

piece; the blade section and cutting edge in another. The latter, of high-quality steel, was welded to the blade back creating a broad, curved section which was markedly broad and wide at the junction point. This facilitated ample thickness for regular sharpening, which thus avoided loss of shape, or form, of the cutting edge. The blade width between this vital joint and the backing section is astonishingly thin. This naturally contributes to the surprising lightness of the head. 'The metal of the blade is remarkably thin, save immediately behind the cutting edge where it spreads to give weight and thickness.' I arranged for the weighing of three axe-heads, the details of which are given overleaf, to judge the effects of the thin blades. The deliberate reduction in the axe-head's weight that was achieved during blade manufacture naturally diminished the warrior's fatigue in combat and increased the variety of blows the axe could effect.

A housecarl. The blade of his axe measured as much as a foot across, with a 5-foot helve. It was held in a left-handed grip to strike a foe's unshielded right side. (Drawing by W F N Watson. From Volume 1 of Battles in Britain *by William Seymour, published by Sidgwick & Jackson, London)*

Occasionally, the sockets were lined with brass rings, some of which were decorated. These also protected the top haft section from sword blows. Some of the blade edges project further forward at the top than at the bottom: a design which facilitated achievement of deeper and longer wounds.

The wooden hafts can be estimated at between 4 and 5 feet long, depending on the warrior's height. The housecarls who wielded them were usually of Danish stock and were predominantly taller and more robust than most other warriors. I have always considered (but not yet authenticated) that hafts were possibly slightly curved in a flattened 'S' form. Sadly, research of the Bayeux Tapestry does not seem to clarify this point. The tapestry reveals some slightly bent ones but the majority are straight. Nonetheless, after many years of research into Anglo-Saxon warfare, I think it probable they were curved because such a shape increases the number of different types of stroke that an axe can deliver, particularly at short range, and much greater power can be generated in close-quarter combat. A long, straight haft does achieve very strong blows at maximum range but is very clumsy to employ at shorter distances.

WFN WATSON

Top sections of some hafts (near the axe-head) may have been protected by metalwork but this does not seem to have been a common occurrence. Certainly a few had decorated metal sockets. It may also be possible that the haft was notched, enabling a warrior to immediately grasp the correct handhold position in relation to his distance from an opponent. Again, this has not been authenticated but may be a credible theory. By the 14th century, however, hafts were generally strengthened with langets comprising strips of metal secured with rivets to the haft.

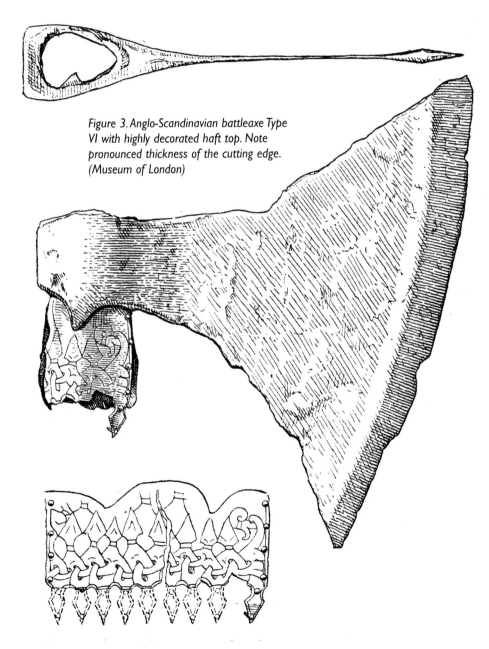

Figure 3. Anglo-Scandinavian battleaxe Type VI with highly decorated haft top. Note pronounced thickness of the cutting edge. (Museum of London)

Specifications

Below are details of three high-quality axe-heads of the Type VI used in the 11th century. These were classified at the Museum of London.

No 1 (A15675)

Height from blade centre to socket top:	206mm
Blade width measured along the curve:	192mm
Blade thickness at joint:	7mm
Blade thickness at centre:	4.75mm
Weight:	749g (1lb 11oz)

No 2 (A2123)

Height from blade centre to socket top:	209mm
Blade width measured along the curve:	234mm
Blade thickness at joint:	5mm
Blade thickness at centre:	5mm
Weight:	438g (15oz)

No 3 (94192)

Height from blade centre to haft top:	214mm
Blade width measured along the curve:	251mm
Blade thickness at joint:	7mm
Blade thickness at centre:	5mm
Weight:	489g (1lb 1.5oz)

Considerable erosion of numbers 2 and 3 has naturally reduced the weights of the axe-heads. Note that number 3 has the longest blade edge. It is difficult to estimate the weight of an uncorroded axe but this could be estimated at about two to two and a half pounds. Some axe-heads with a blade edge of about one foot have been discovered.

MILITARY EFFECTIVENESS

In the 11th century the English regular army (paid by the nation) comprised about 3000 housecarls. These were carefully selected men of large stature (6 feet or taller), impressive physical fitness and very competent in handling the sword, spear and battleaxe. They were the most feared infantry in Europe because of their courage, close teamwork, and efficiency in combat. Their battleaxe was a particularly decisive in close combat, the area in which most of the fighting took place in contemporary battles. Warriors were taught to wield their axes over the left shoulder to avoid an opponent's shield. This is confirmed by the housecarls' success in the Welsh campaigns and at the battles of Stamford Bridge and Hastings. During the latter the battleaxe inflicted significant casualties, particularly of warhorses. 'The broad axe of the 11th century was a fearsome weapon of immense force when wielded with strong arms.'[2] It possessed one disadvantage: when the axe was held with both hands the shield could not be carried to protect the warrior and, naturally, such a large weapon was always used in two hands. This problem was overcome to an extent by lightly digging the shield into the ground, which gave a secure base when not using the axe. Housecarls usually fought in pairs, particularly when confronting cavalry. In battle, a warhorse occupied space equating to that held by two foot-soldiers. The one on the left, using his sword or spear, would employ his shield to deflect the mounted knight's spear or sword. This enabled the one on the right to use his axe. While the horseman was engaged with his comrade, he would skip forward, swinging his axe in a huge arc, and either remove the knight's leg just below the knee or kill the horse, or, sometimes, kill both in turn. This was a terrifying and highly effective combination, as the Norman French discovered.

We have briefly examined the evolution of the Viking military axe in Britain, and studied the designs of the bearded axe, and the construction of the most famous axe – the 11th century Type VI two-handed battleaxe. This indicates how a technologically advanced design, combined with efficient production techniques, achieved an axe-head of graceful form, considerable strength, good balance, devastating cutting edge, and surprising lightness. This effective type of arm was retained, in modified forms, as a popular weapon for foot-soldiers until the late 15th century. For some Viking and Anglo-Saxon warriors the axe was their primary weapon.

Chapter Nine

Later Anglo-Saxon Swords

This chapter examines briefly sword production in Britain during the later Migration period, and then studies later Anglo-Saxon times, during which so many high-quality and beautiful swords were produced, in the 9th and 10th centuries. This embraces the decorative influence on some English arms of the famous Trewhiddle artistic style. Other matters allied to weapon research, such as the decoration of English swords, the rather strange discovery of many sword examples in isolated waters, continental sword developments, and weapon corrosion are also considered.

Sword Production in Britain

In the very early Migration period the Anglo-Saxons may periodically have been resupplied with new pattern-welded or 'steely iron' blades from the Continent. They might also, perhaps, have received processed iron in the form of bars and blanks. Such procedure could have continued for some time. However, it is also possible that some tribal chieftains brought to Britain their own swordsmiths with the aim of setting up new forges where locally made weapons could be manufactured. This poses the question of how and from where they obtained their iron ore supplies. Much uncertainty remains on this subject. 'If Anglo-Saxon blades were made in England, what was the source of iron, and where were they fabricated?'[1] I consider it possible that at least some old areas of Roman iron ore mining, and iron artefact manufacture, may eventually have been rediscovered and worked to acquire local supplies of iron ore. After all, the Romans established numerous iron mines, particularly in south-east Britain in the Weald area, at such places as Hartfield, Kingscote and Hammerwood, all in the area of East Grinstead, and Broadfield and Holbeanwood. Another example is at Petley Wood in Sussex. 'In the area of Kent and Sussex are two items of documentary evidence for iron working at this period.'[2] A charter of AD 689 mentions an iron mine near Lyminge, Kent, while the Domesday Book lists ironworking 'near East Grindstead' in Sussex. Evidence of migrants using these sites is sparse but, doubtless, more convincing proof will eventually emerge. 'Haslam (1980) suggests that at least during the 8th and 9th centuries changes in technology occurred which point to the rediscovery of processes lost to England since the Roman Age.'[3]

Later Anglo-Saxon and English Swords

'In England true pattern-welded swords continued to be forged by Anglo-Saxon smiths into the late 9th and 10th centuries, and during the earlier period some

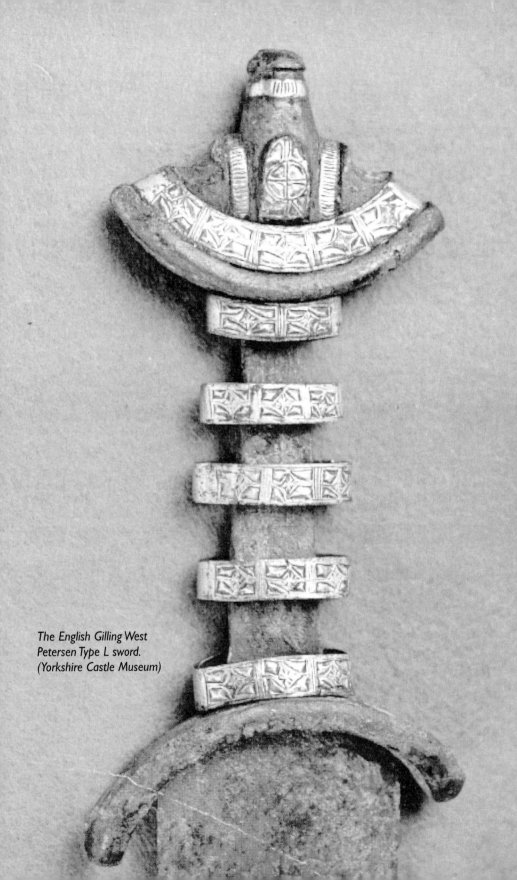

The English Gilling West Petersen Type L sword. (Yorkshire Castle Museum)

The English Wensley pattern-welded Type L sword from a Viking grave in a churchyard in North Yorkshire. Note that the round, curved guards on L-types were undecorated. (© Copyright the British Museum)

were of the Petersen Type L.'[4] This created a marked contrast with the Continent, where new, plain but high-quality steel blades, some with pattern-welded inscriptions, were manufactured from after about AD 850.

In the second half of the 9th century some established English swordsmiths successfully developed an impressive and revolutionary sword-hilt form. It had pattern-welded straight blades which were double-edged and fullered. The pommels were elaborately designed, and were remarkable for their substantial size and beautiful decoration in the Trewhiddle style (see page 115). The central pommel lobe was fashioned in a high coned shape (see picture on page 109). There is a detailed description and illustrations of the Gilling West sword in Chapter 10. This arm is a Petersen Type L, discovered with others in North Yorkshire.

The upper pommel guard on these was now sharply upward-curved in a very conspicuous manner. The lower main guard, which traditionally had been straight, was now rounded, made in undecorated iron and was of radically different design to any other of the Migration period, being very sharply curved downwards in a neat semi-circle. This feature provided an effective means of trapping an opponent's blade. On some examples the guard, quillons extended further down the sides of the blade but all of the guards provided an immediately recognizable feature. Examples of Petersen's Type L did sometimes differ slightly but all retained common features of high central pommel sections, sharply upward-curved upper pommel guard, and acutely downward-curved lower guard. On the Gilling sword the hand grip was further secured and enhanced with decorated solid-silver bands which demonstrated further status and provided a better hand grip. The novel features of the Type L sword were entirely of English origin. 'The L Type occurs in the British Isles and Norway, and even once in Iceland, but as the ornamentation found on its silver plates is in the Trewhiddle style, there is no doubt of its origin in England.'[5]

We now examine another very fine later English sword, which was sometimes misinterpreted in the past as a Type L. This is the Abingdon sword (see picture overleaf), which harbours several differences with the L Type. According to Vera Evison:

> 'The pommel was entirely covered by a silver cap instead of by separate inlaid plates, and the guards are very different from the L Type. Instead of being thin with a convex face they have become broader with a flat surface which now provides space for decorative silver plates. The guards have also increased in length, the upper guard extending beyond the breadth of the pommel and the lower guard to a length of 12.5cm. The Abingdon sword therefore provides a link between the L Type (Gilling West sword) and the English Wallingford Bridge type.'[6]

The latter was found in the Thames in Berkshire.

> 'While the Wallingford Bridge sword has a general similarity to the L Type in its curved guards and tri-lobed pommel, there are dissimilarities. The middle lobe of the pommel is the largest of the three and is rounded rather than cone-shaped as often on the L Type, the reason for this difference being that the pommel of the L Type is fixed on the tang itself which passes through the

The magnificent 10th C. Abingdon sword. Note the differences between this and the Type L ones. This outstanding sword provided link between the L- types and the later Wallingford Bridge sword depicted opposite. (The Ashmolean Museum, Oxford)

pommel. The lower guard is longer, and instead of having the top view of an oval with nearly parallel sides and rounded ends as on the L Type, it now has plates with the animal, foliate or geometric patterns of the Trewhiddle style, whereas the Wallingford Bridge type is never embellished with silver plates of this type. The Wallingford Bridge type of sword, then, represents a typological development between the L Type and the sword types of the 12th century with long and slender curved guards. Its inception period is securely fixed, as the last half of the 9th century is the period that saw the creation of the L Type, with its Trewhiddle-style decoration. The "sword of no provenance" in the British Museum (BM. Tr.173) and the Abingdon sword point the way to the broader and longer guards of the series of swords with inlaid metal strip decoration, mostly from the Thames, which must belong to the 10th century. Between these and the 12th century types must be placed the longer and more elegant guards of the Brentford, Windsor, Battersea, the Thames 1840 and Wallingford Bridge swords. The imported blades associated with these hilts also develop from pattern-welding to "Ulfberht" and "Ingelrii" blades, then to imitations of these inscriptions and marks only. To archaeological evidence may be added the art historical, which postulates the Winchester style period. Pictures in manuscripts point certainly to the 10th century, and probably to the first half of the 11th also. All sources therefore agree, but without the present possibility of further precision. However, the discovery of the Wallingford Bridge sword and the revealing decoration on the Thames 1840 hilt must increase our respect for the Anglo-Saxon sword-smith.'[7]

Wallingford Bridge sword 'representing the typological development between the 'L'Type and swords with long and slender curved guards of the 12th C.' (Vera I Evison). (Reading Museum Service, negative number 1965. 170. 1.)

Vera Evison further suggests that the latter sword, found in1840, was very similar to the Wallingford Bridge one (see figure on page 115) and was probably made by the same smith.

'Another sword, no doubt also by the same hand, with a different-shaped pommel but the same type of guard and ornamentation was found in the Thames at Battersea. These swords belong to a group which can now be recognised as of Anglo-Saxon manufacture, most of them having been found in the Thames.'[8]

The Shifford sword from the River Thames above Tenfoot Bridge. Decoration 'on a background of vertical and diagonal striations in the iron was laid silver plate, and copper strips were inlaid in a running lozenge pattern with a double copper tendril motive inside each lozenge.' V I Evison. (Reading Museum Service, negative number 1947.285.1)

Decoration of late Anglo-Saxon (English) Swords

As I mentioned, from the mid-9th century to the 10th century the metal hilt sections of many Anglo-Saxon (English) swords were decorated in the arresting manner of the Trewhiddle style. This style may have continued later in northern England although its name originated with the discovery in 1774 of a precious hoard in an old mine at Trewhiddle in Cornwall. Among the items recovered were some animal-ornamented mounts from a drinking horn, and two silver chalices. The fashioning of the decorative panels involved the employment of costly materials such as very fine gold, pure silver and intricate niello panels. The hoard's deposition has been dated as about 868 from a coin that was found there. According to Ellis Davidson in *The Sword in Anglo-Saxon England*, 'the Trewhiddle style to which T.D. Kendrick in his *Anglo-Saxon Art*, London 1938, has given the name of "West Saxon baroque" and is found on metalwork of the Trewhiddle hoard.'

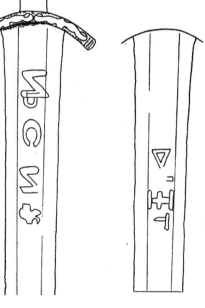

Late Saxon sword from the Thames found in 1840, now in the Carlisle museum. Note its beautiful decoration. Professor Evison believes that this English arm and the Wallingford Bridge one were probably made by the same smith. (Carlisle Museum, and Cumberland and Westmorland Antiquarian & Archaeological Society. Drawing by courtesy of V I Evison)

'Those (swords) decorated in the Trewhiddle style, for instance, are found in England and Norway, but are typically Anglo-Saxon in choice of motive and ornamental treatment, and probably came from one workshop.'[9]

'The effective ornamentation in animal and foliage designs is characteristic of what we know of Anglo-Saxon decoration on metalwork in the late 8th and 9th centuries.'[10]

Vera Evison states:

'the inlaid work on the hilts of the Battersea, Wallingford Bridge and the Carlisle Museum sword (recovered from the Thames in 1840) is similar in execution and the basketwork pattern on the top of the guards is used on all three. The acanthus pattern on two of them suggests a 10th to 11th century date, and this is supported by the form of the hilts and contemporary manuscript illustrations.'

This late 10th to 11th C. English sword with imported blade from the Rhineland interlaid with 'Ingelrii' in iron Roman letters, was recovered in two pieces from the Thames at Wandsworth. It is possible the very-well-known title 'Ingelrii' was actually used as a trade name by a much later smith. The English hilt has pommel with central globular pommel with two zoomorphic flanking lobes. These are segregated by vertical lines. The guard is the same pattern as the Exeter one. The sword is Wheeler's Type VI, Petersen's Type Z. (Museum of London)

Sword Inscriptions

According to Janet Lang and Barry Ager in their paper '*Swords of the Anglo-Saxon and Viking Periods in the British Museum*', 'Inscriptions on sword blades appear to have been made by hammering short lengths of plain or twisted wires into a chiselled channel in the blade surface whilst hot. The characters were secured in place by further hammering after reheating.'[11] Doubtless a punch was employed to move the letter inlays into the correct and level positions. Thereafter, blades were ground and burnished to achieve a contrast between the pattern-welded inlay and the darker blade. They mention that Modin (unpublished) discovered that the pattern-welded letters on Ulfberht blades were made of ferritic iron which might have contained phosphorus. This would probably have remained bright and shiny against a dull background. After this exciting discovery an increasing number of steel sword blades were made on the Continent bearing the names of continental smiths or their workshops. We cannot be sure whether all the workshops using these names were under the control of certain smiths or merely copied the name as a mark of quality. Certainly, the long use of these names rules out the direct personal involvement of the persons concerned during this later period. They included well-known names such as Ulfberht, Ingelrii and Leutlrit. Some of these imported blades were fitted with the beautiful and practical English hilts described above.

The copper-alloy Exeter lower sword guard of 10th to 11th C. Made by an Anglo-Saxon (English) craftsman. Inscribed (L)EOFRI(C)ME F(ecit) on top. Different designs on both sides: key pattern on one, geometrical interlace on other. Length: 3.3in (8.4cm). This later, popular guard design was used on several arms including the Shifford, Brentford and Wandsworth swords. (© Copyright the British Museum)

CONTINENTAL SWORD DEVELOPMENTS

We now briefly examine technological developments on the Continent from the late 8th until the early 10th century. In the Rhineland a thriving sword-manufacturing industry had gradually developed over the centuries, primarily due to rediscovered sources of high-quality iron ore. Additionally, generations of accumulated technical knowledge and expertise of metallurgical and forging techniques guaranteed the production of the highest-quality sword blades. Finally, more effective and efficient forges must surely have been introduced. It was, of course, in the Rhineland that better-balanced blades with a higher blade centre of gravity were introduced (see Chapter 7), which markedly improved sword handling and increased the variety of strokes that could be employed. It was also in this area that the famous pattern-welding system of sword production was gradually replaced by the making of high-quality steel blades. This resulted in a much faster, and perhaps cheaper, sword production rate. The ancient process was, however, retained for sword-blade inscriptions and veneers. This indicates the continued social importance to warriors of owning a blade with an intriguing and attractive appearance.

Chapter 7 introduced the Types VIII and IX swords, which were added by Ewart Oakeshott to Wheeler's typology of Viking sword-hilt forms to span the period between the later Viking period and the 11th to mid-12th century. Many swords of this period generally retained a form of blade that had been popular in the Viking period.

Various new pommel forms, such as the Brazil nut and the tea cosy, were used on the hilt of the Type VIII, and the cocked-hat and wheel pommels were employed on the Type IX's hilt. However, pommel selection did vary, and other forms were also used. Eventually some later blades were narrower and slightly longer than Viking ones. Their guards were also longer: about 7 to 8 inches, rather square-shaped and narrower than previously. These were generally straight though occasionally slightly downward-curved at the ends. Presumably, these deflected an opponent's sword blows more effectively, particularly when delivered by a mounted warrior. The narrower, longer, more pointed blades could also penetrate mail better, when delivered in a straight thrust. Some men retained Viking or Rhineland blades fitted with a range of these hilt artefacts. Overall, this was very much a transitional period, when sword components of various periods were amalgamated into one weapon.

Sword Burials and 'Lost' Swords

In the radiographic study undertaken at the British Museum by Janet Lang and Barry Ager on 142 Anglo-Saxon and Viking swords it was noted none was dated to the 8th century. The reasons for this interesting and initially surprising revelation are still rather obscure. Perhaps it is just a coincidence. Perhaps during some periods of that century there were major hostilities, which caused weapon shortages and thus the decline of arms interment. Was the cause related to Christianity? By the 8th century many of the Anglo-Saxon population had been converted to Christianity. However, Ager says:

'There is no evidence at this period that the church proscribed burial with grave goods and even ecclesiastics such as St Cuthbert and members of the elite were in fact still buried with them. The transition was clearly a complex process and extended over quite a long period. Current opinion is that priests only gradually gained control over burial arrangements as churchyards became established and that it was rather the introduction of Christian practices than the putative and undocumented prohibition of pagan ones that brought about a change. Some would go further and say that in fact the Church's role in this was relatively minor and that the transition was more due to social changes.'[12]

Whatever the reason, the lack of interment artefacts does deprive archaeologists of a fruitful source of information. Fortunately, later invasions of pagan Viking warriors to England caused pagan burial practices to recommence, to some extent at least, in the areas they managed to control.

Providentially, however, weapons continued to be lost in rivers and at battle sites, and these are periodically discovered. Ellis Davidson commented:

> 'There are a number of fine weapons in English and continental museums which have been dredged up from rivers, many of them dating from the Viking period. It is generally assumed that these were lost in battle, or dropped at a ford, but the possibility of a deliberate sacrifice, as at Illerup, must be borne in mind.'[13]

River-crossing places, particularly at fords and bridges, were often of tactical importance during the Dark Ages as these tended to mark territorial boundaries. Therefore, battles or affrays occurred there, sometimes causing weapon losses. River or lake finds of 8th century arms far away from possible battle locations could indicate a reversion to the pagan custom of casting a rare and valuable sword on waters considered sacred. It is thus possible that some of the pagan warriors, who had only recently converted to Christianity, may have considered it advisable to hedge their bets to guarantee their salvation. Such men might have agreed to be buried in accordance with Christian procedures but also stipulated to close relatives that after death their sword should be deposited in accordance with ancient pagan rituals, and with due reverence, in an enchanted, magical section of a river or lake location. This could conceivably explain the periodic discoveries of swords in secluded places.

Corrosion of Weapons

It is interesting to consider the effects of water and soil on weapons recovered from these elements. Grave swords buried in soil, particularly of acid nature, can become very corroded, while those from normal soil, or water, may be better preserved. I examined a Migration period sword (E Behmer Type IV) in the reserve arms collection of the Salisbury & South Wiltshire Museum. This was in fine condition although then unconserved. It had been discovered in a grave, which must have been a very chalky one, in a pagan Migration period cemetery in Salt Lane, Winterbourne Gunner, in Wiltshire. It lacked all sword items except the blade and tang. However, these were in a remarkable state of preservation, which must have been predominantly due to the chalky earth in that region. An indication of this is confirmed by its weight of 2 pounds 6 ounces.

Swords were habitually buried in their scabbards. Sometimes these were further wrapped in linen clothes. Their purpose was to provide maximum protection for the valuable arms. 'Some scabbards were probably covered outside with linen over the leather, as in later medieval times; the Lavoye scabbard dated the 6th century, had a leather covering under a second cover of stout linen.'[14] 'Swords taken from graves have frequently rusted in their scabbards, so that it is impossible to examine the blade.'[15] This is true – I have examined only about three weapons on which the majority of the scabbard remained. Generally, most scabbards have disintegrated almost completely, sometimes leaving only tiny fragments of wood and leather on the blade, particularly beneath the upper scabbard plate and chape. It has often been noted that one of the first blade areas to suffer corrosion is at the blade tops, where moisture could eventually penetrate the scabbard mouth. Weapons in water seem to suffer less corrosion damage. 'A rather surprising result of immersion for centuries is that in general a sword taken from water is in better condition when recovered than one which has been buried.'[16] When working in the Museum of London I handled many arms that had been recovered from the Thames. Several were encased in clay and little pebbles, which provided a rather effective sealing preservative. Weapons that had lain in the mud at the bottom of a river often revealed one blade side, the lower, in much better condition than the other.

Chapter Ten

Four Notable Swords in Britain

This chapter examines in close detail four swords of special significance and interest which fit into the following Type groupings: two are from the Germanic Migration period, one with a parallel-sided blade, the other with a tapering blade; a beautifully decorated English sword, and a 10th century Viking one. The detailed descriptions of these will enhance the more general information given in the previous chapters.

GERMANIC MIGRATION PERIOD BROAD-BLADED SWORD AT DEVIZES MUSEUM

This very fine, substantial and impressive double-edged arm is made from forged iron and some bronze. It is accession number 'grave 22' and was recovered from the Pewsey cemetery near Devizes. The sword belongs to the Blacknall Field archive number 1991.1. Its X-ray number is: 750130. The weapon was pattern-welded and primarily designed to achieve powerful hacking or cutting strokes delivered from above the shoulder.

The grave from which this was recovered was obviously a very rich one, containing numerous artefacts such as an iron shield-boss, iron studs which retained the boss to the wooden shield, the iron shield-grip, a bronze link attachment, the iron sword chape, circular bronze bucket-strips, socketed iron spearhead incorporating its ferrule, an iron knife, an iron buckle and a silver gilt bronze buckle. The sword's iron U-shaped chape is of fairly simple construction, being hollow and semi-circular in form, designed to securely retain the bottom sections of scabbard laths and their leather coverings. The front section is higher than the rear to protect the scabbard from ground foliage. This design form differs from the majority of chapes, which were of the

Broad-bladed sword from Devizes. A highly significant and rare upper scabbard plate depicting scroll borders, below which are two zoomorphic dogs, or possibly serpents. The whole artefact would have been gilded. This plate is of unusually high quality, substantial, deep and wide, being 7.7cm (3.03in) and 2.4cm (0.945in) high. Note the wooden scabbard remnants on the blade. (Devizes Museum)

same height on both scabbard sides. The weapon can be dated between about AD 500–550. It was accompanied by one glass amulet bead and several metal sections relating to the baldric system. Considering its age, the condition is very good, although the scabbard, hilt grip and guard are missing. The loss of these components is, sadly, very common. Naturally the long corrosion period would first attack the leather and wood scabbard and then the horn or bone hilt components and cross guard. However, the blade shows considerable signs of scabbard wood and leather fragments, particularly beneath the upper scabbard plate. The pommel cap is detached from the upper pommel guard plate (bar) set on the tang tip. This is of neat, 6th century form, and is 2.2 inches wide, 0.6 inches high, and hollow. It is accompanied by its two bronze (gilded) retaining pins.

The blade, which was constructed by the pattern-welded method, is substantial. Its measurements are:

Overall length:	35.60in (905mm)
Blade length:	30.79in (782mm)
Blade width at top:	2.30in (60mm)
Intermediate:	2.25in (57mm)
Blade width at centre:	2.10in (51mm)
Intermediate:	1.88in (45mm)
Width above point-section:	1.65in (42mm)

I consider the sword to be of the Elis Behmer Type IV because the blade edges appear almost parallel, a classification further supported by the blade widths, which compare to the usual width of this type, which are described as varying from about 1.75 to 2.3 inches. This comment on sword widths is interesting: 'Swords from the cemetery at Sarre (Anglo-Saxon) vary from 34 to 37 inches (86.4 to 94cm), and there is plenty of minor variation, since no two swords from this cemetery are precisely the same length.'[1] So our sword blade, at 30.79 inches long (782mm), falls into the upper band of this range.

The blade was apparently not fullered, and the X-ray confirms this. However, because this shallow feature was one of the first sections of a sword to be corroded, we cannot be certain that it did not originally exist.

The upper scabbard plate is exceptionally fine and rare. It is deep and large, and is finely and elaborately finished in a moulded bronze alloy which was subsequently gilded. The decorative features are beautifully and delightfully executed, and depict a scroll border and an upside-down man flanked by two zoomorphic animals resembling large dogs looking backwards over their shoulders. The decoration is very fine and makes the artefact of considerable significance.

Many metal sections that were incorporated on the scabbard, including buckles, survived with the sword. All these related to the system by which the weapon was attached to the waist, thus enabling the warrior to carry his sword. The baldric arrangement, which was fairly sophisticated and efficient, was embellished by decoration of some features. Among these were bronze chain links attached to rings, and two decorated, rounded, gilded bronze sections 3.45 inches (8.7cm) long fitted to top scabbard sides, through which the belt passed. A buckle-strap holder was mounted some 13 inches down the scabbard rear. The strap from this buckle extended back to the rear of the waist belt, where it was connected, perhaps with another buckle. This was designed to keep the sword and scabbard set at a backward angle keeping the scabbard-bottom above ground foliage. It also ensured that the sword hilt was set at a convenient angle for the weapon to be drawn.

This fine weapon, with its highly important upper scabbard plate and large overall proportions, is of great significance and consequence. Without doubt, its possession would have conferred considerable status on the owner who, although perhaps not a king, must probably have been at least a tribal leader of consequence.

GERMANIC MIGRATION PERIOD TAPERING SWORD AT CROYDON MUSEUM

This double-edged tapering sword of the Saxon/Frankish Migration period was recovered from the cemetery at Elderidge Road, Croydon in 1893/8. The rather small but graceful sword, with its very gently tapering blade, is retained at the Croydon museum. Its accession number is M1992/10 and its estimated age from about AD 550–600. The following items are missing: the pommel cap, grip, lower pommel guard, most of the scabbard, and the rear of the upper scabbard plate. The condition is generally fair but the chape is in a good state.

The blade, which is very well designed and fashioned, was made in the pattern-welded method. It is rather less clumsy than the parallel-sided type such as the Devizes one described above because its blade is narrower. It could inflict fairly powerful over-shoulder hacking strokes but was probably primarily designed to effect straight, thrusting lunges at an opponent's body. We can deduce that it was easier to handle than Type IV swords and thus would have enabled a slightly greater variety of combat strokes to be effected. Note that, owing to the chape presence at the blade bottom, the final measurement could not be made precisely. However, the X-ray data shows that the blade tip is fairly pointed. The rectangular, gilded bronze alloy upper scabbard plate is decorated with horizontal grooves, and is very similar to plates found on swords from the Mitcham Anglo-Saxon cemetery. These may thus have been made by the same craftsman. The plate is 2.5 inches wide (62mm) and 0.4 of an inch deep (10mm). Unfortunately the rear plate section and the retaining pins were missing (see picture overleaf). Most of the scabbard sections are also missing but deposits of wood and leather still remain on the blade. Significant sections remain under the upper scabbard plate and between the chape sides.

The sword measurements, which demonstrate the fine tapering blade, are as follows:

Overall length:	32.90in (836mm)
Blade length:	28.20in (709mm)
Blade width at top:	2.00in (50mm)
Intermediate:	1.95in (48mm)
Blade width at centre:	1.79in (47mm)
Intermediate:	1.75in (44mm)
Blade width above point:	1.50in (37mm)[*]

[*] (Estimated owing to chape and leather fragments obscuring blade).

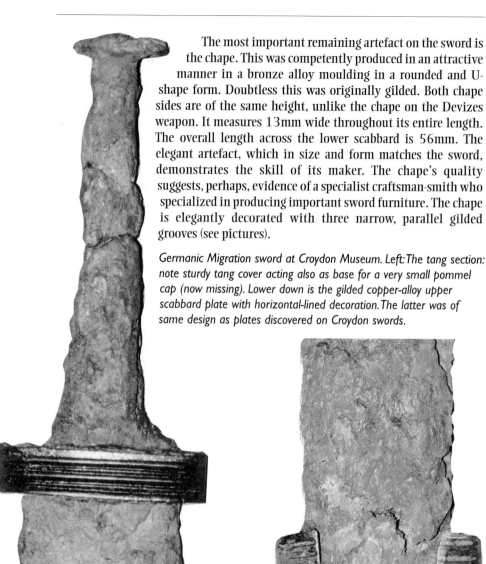

The most important remaining artefact on the sword is the chape. This was competently produced in an attractive manner in a bronze alloy moulding in a rounded and U-shape form. Doubtless this was originally gilded. Both chape sides are of the same height, unlike the chape on the Devizes weapon. It measures 13mm wide throughout its entire length. The overall length across the lower scabbard is 56mm. The elegant artefact, which in size and form matches the sword, demonstrates the skill of its maker. The chape's quality suggests, perhaps, evidence of a specialist craftsman-smith who specialized in producing important sword furniture. The chape is elegantly decorated with three narrow, parallel gilded grooves (see pictures).

Germanic Migration sword at Croydon Museum. Left: The tang section: note sturdy tang cover acting also as base for a very small pommel cap (now missing). Lower down is the gilded copper-alloy upper scabbard plate with horizontal-lined decoration. The latter was of same design as plates discovered on Croydon swords.

Right: The fine bronze-alloy iron chape is decorated with short parallel grooves, some still retaining gilding. Note the right rivet has lost its cap. Between the chape sides are major remains of the leather and wood scabbard. Above this on the blade can be seen additional major scabbard remnants. (Both reproduced by kind permission of the Croydon Museum & Heritage Service)

FRANKISH VIKING SWORD OF ABOUT AD 900–950 IN THE WALLACE COLLECTION

Historians researching the middle period of the Viking expansionist movement will from time to time be fortunate enough to find a particularly rare and fine sword. The Wallace Collection possesses just such an arm. This typifies one of the best-quality arms sometimes carried by the Vikings and confirms the metallurgical and technological skills of the 9th and 10th century Rhineland swordsmiths. The weapon carries accession number A 456 and is on display in the Wallace Collection European gallery. It is classified by Petersen as Type K and Mortimer Wheeler as Type IV.

Despite much wear, evidence of active blade-honing and, perhaps, the effects of river immersion, the sword remains a very fine one. It seems more likely to have been recovered from a river than from the ground because the latter would probably have caused considerably more corrosion, and the weapon, made of high-quality steel, has an attractive patina. The blade is of classic Viking contemporary form, double-edged and tapering gently to half an inch above the point, with a fuller on both sides of the blade. Thanks to the kind co-operation of the Museum of London and the specialist expertise of Brian Gilmour of the Royal Armouries, the weapon was given an X-ray examination. This confirmed that: the pommel was definitely constructed in two sections; the blade was not pattern-welded, and that the blade has an inscription on one side, 1.5 inches (3.8cm) below the guard, comprising either a letter 'C' or a symbol similar to an incomplete omega.

The sword measurements are:

Overall length:	35.7in (90.6mm)
Blade length:	30.2in (76.5mm)
Guard length:	4.0in (10.2mm)
Grip length:	3.48in (57mm)
Blade width at centre:	2.1in (5.2mm)

There are broad, shallow fullers about 0.75 inch wide on both sides of the blade which stretch down to about two inches above the point. The cutting edges, although showing wear from sharpening, are still in good condition. The weapon handles very well, which is to be expected of one with a high balance point. It lifts easily from the ground to above the shoulder and from there easily to a good position to protect the body. The sword would have been an excellent combat arm capable of delivering straight thrusts and over-shoulder cutting strokes. The pommel is of the classic 'cocked-hat' form, comprising two sections (see picture). The top one incorporates five lobes, the central one of which is slightly higher than the others. The lower section is the pommel bar, which provides the base to the top section. This is pierced at both lower ends to hold rivets that helped to keep the two sections together. Some signs of the brass decoration, described below, are still visible.

The iron cross guard is typically rather modest in size, and is rather boat-shaped in form with rounded ends (see picture). Small, round holes drilled through the guard are sited at both ends. These would have held decorated pins above and below the guard, which are now missing. Perhaps they were removed at some time by a looter who erroneously assumed that they were gold. The purposes of these were possibly decorative but also as a vestigial reminder of the ones employed during the Migration period for the practical purpose of retaining the sandwich components of the guard. The guard is marked with fine, very closely set vertical incisions. These provided the base for the subsequent addition of a brass coating, which was hammered into the parallel cuts until the whole surface was covered, thus providing attractive decoration. Both sides of the cross guard were exquisitely executed with a vine-branch decoration achieved by incising through the brass and revealing the iron beneath. No grapes were in evidence with the vine. This pattern can still be seen fairly easily.

The sword has a further significant attribute. It is one of a group of five weapons that are inscribed on the cross guard. Whether these names relate to the maker of the sword or just to the hilt components and their decoration is uncertain. It is probably the name of the craftsman decorator and not the sword owner. Some of the five names are spelt differently, which is not surprising as the craftsman was doubtless illiterate. However, the similarities in the decoration on the five swords are such that they were most probably fashioned by the same craftsman. The studies on this decoration undertaken by Ewart Oakeshott and Ian Peirce and published in the 12th Park Lane Arms Fair catalogue are of great interest. The other four swords in this group were: one found at Kilmainham marked 'Hartolfr' but only the 'H' is now discernible; one from Ballinderry marked 'Hiltipreht' on the guard and 'Ulfberht' on the blade, and grapes were also included in the decoration, and one sword originally on display in Berlin, but destroyed during the war, which according to F Morowe was marked 'Hiltipreht' and also had grape decoration. The final one, (the Malhus), which also has grapes in its decoration, is in the Trondheim museum, and is marked 'Hiltipreht' on guard and has a pattern-welded inscription of 'Ulfbertht' on the blade. On all but the Wallace example, the precious metal employed in the decoration was silver.

Very fine Frankish-made Viking sword of Petersen Type K and Wheeler Type IV, accession number A 456 in The Wallace Collection. (Reproduced by Permission of the Trustees of the Wallace Collection)

View of boat-shaped guard-top revealing the three letters 'HIL', presumably of the name 'Hiltipreht'. Sadly, other letters have been lost through corrosion. Note the very closely set incisions on guard providing efficient base for later application of brass. (Reproduced by Permission of the Trustees of the Wallace Collection)

The Wallace Collection sword guard only has three letters visible: 'HIL'. Brass was used in the decoration instead of silver, and no grapes are included. According to Ewart Oakeshott, there is a final common feature on three of the swords. These are 'rabbits' ears' sited beneath the cross on the Wallace Collection, Ballinderry and Malhus swords. The evidence on the last is now only faint.

Despite the loss of its rivets, wear to its decoration and the absence of its scabbard, the Wallace Collection sword remains a very fine and emotive weapon. It still exudes, in a compelling and strangely mesmeric manner, a sense of its past status and significance.

THE GILLING ANGLO-SAXON (ENGLISH) SWORD PRE-AD 866 IN THE YORKSHIRE MUSEUM

This weapon is a beautiful example of a late 9th century Anglo-Saxon sword which we can accurately describe as English. The decoration, examined below, incorporated panels containing plant and other designs typical of the English Trewhiddle style. We can thus deduce, this arm and others of very similar form were fashioned early in the most noteworthy period of the high Anglo-Saxon artistic Renaissance. I have known and greatly admired this piece for many years. Ellis Davidson, in *The Sword in Anglo-Saxon England*, states: 'Wheeler calls the acutely curved hilt the Wallingford type, after the well-known hilt in the Ashmolean Museum which was thought until recently to have been found at Wallingford, but is now known to have come from Abingdon.'[2] Our weapon is a classic Type L as categorized by Petersen. This sword was designed, made and decorated by Anglo-Saxon craftsmen; the sword type is earlier than the Abingdon sword. To assist in research on the Gilling sword I studied Brian Gilmour's academic paper 'A Late Anglo-Saxon Sword from Gilling West, N. Yorkshire' published in *Medieval Archaeology*, vol. XXX, 1986 by the Society of Medieval Archaeology. This states that the weapon was found in April 1976 by a nine-year-old boy in the Gilling beck, near the village of Gilling West, in North Yorkshire – a fortuitous discovery. The actual find site was some 33 feet east of a bridge and at a point where the Gilling beck is open at both sides of the bank. Furthermore, the beck's bottom here is hard-cored and rutted, and was therefore very possibly a ford. Such locations were, during the Dark Ages and Anglo-Saxon times, places of military tactical significance and were occasionally the scene of affrays or little battles. Consequently, weapons are regularly found in such places. Evidently, the area of the beck in which the arm was discovered had been previously cleaned and widened by an excavator. This activity may have caused the movement of the sword from its original location. 'Consequently the context of the sword's deposition must be regarded as uncertain, and one possibility is that it came from a Viking grave.'[3] This is plausible because two others of very similar form were recovered in North Yorkshire from Viking graves at Camphill and Wensley. They are now at the British Museum and bear very close resemblance to the Gilling sword. Their condition, however, is much more corroded, as one would expect from grave finds. The Gilling sword, conversely, is in very good state, strongly indicating that it had been deposited in and then preserved by water and silt, in this case the Gilling beck. This could indicate that the sword may have been dropped at this location during a little battle.

This sword is on display in the Yorkshire Museum with accession number YORYM 1979. 51. It has a double-edged, fullered blade and an elegantly and richly decorated hilt. The pattern-welded blade is 700mm (27.56in) long, and the overall length 838mm (33in). The blade width just below the guard is 86mm (3.38in). There is a full description of this noteworthy blade by the eminent metallurgical expert Brian J Gilmour on page 129. This includes an enlightening technical account of how its high-quality pattern-welding process was carried out.

The hilt is the sword's most distinctive area and from it the weapon type can immediately be recognized as a high-quality Anglo-Saxon (English) one. The lower, undecorated guard is acutely curved downwards towards the blade edges. This must have been a very effective means of halting then, possibly, holding an opponent's sword slash or thrust. This particularly efficient form of hand-protection is of great interest, as it was very different from those of previous centuries, consisting of modest, straight cross guards. It is intriguing that the English swordsmiths designed and introduced such a revolutionary and efficient form of defence.

The pommel is even more distinctive and unusual. The upper guard plate curves most sharply upwards, away from the blade, and in the opposite direction to the lower guard. This area is not decorated. Above it is set the large and unusual trilobate pommel, which is reminiscent of an ice-cream cone. This is decorated in a complex but captivating manner that clearly demonstrates the very high standards of Anglo-Saxon art in this period (see Figure 3 on page 133). For example,

'On each side of the middle lobe of the pommel, between the narrow silver bands, is a convex-topped rectangular silver plate. Each plate is decorated with a circle divided into quadrants, the circle bordered by four triangular fields below and two sub-triangular fields above. A narrow silver band decorated with two groups of incised parallel lines encircles the middle lobe near its terminal. A silver band follows the curve of the pommel guard, each side decorated with five sub-rectangular panels each containing conventionalised plant ornament in a concave-sided square creating four sub-triangular fields.'[4]

The decorations are enhanced by the effective use of niello. Note that the rounded lower guard was not decorated.

Around the grip (now missing) are five silver bands, which initially encircled the grip. These demonstrated considerable status while providing a firmer hand grip for the warrior. The top one has decoration in two panels; the remainder have the same ornament in three panels. A feature of great interest concerns the central band, which has scuff-marks on it. These are thought to be the result of a glancing blow, which, if the supposition were correct, would have penetrated the man's hand. Did this force the warrior to drop his sword in the beck? It is a fascinating conjecture. The Gilling sword is an outstanding weapon, by virtue of its very-high-quality pattern-welded blade, and magnificent and rich decoration, which amply demonstrates the high standards achieved by English swordsmiths, and the excellent skills of the jeweller who embellished it.

THE GILLING SWORD BLADE

By B J Gilmour, reproduced with the kind permission of the Editor, Professor John Hines, of *Medieval Archaeology*, vol. XXX, 1986, published by the Society of Medieval Archaeology. I also wish to thank Brian Gilmour for kindly allowing the reproduction of his excellent technical paper on the Gilling sword.

The blade is 700mm long from the lower guard to the tip. Its main part tapers gradually away from the lower guard where its width is 52mm. The waterlogged conditions in which it was found have preserved the blade very well, especially on one side where much of the original surface survives as a hard, shiny black patina. A shallow fuller runs down the centre of the blade, on either side, and along this fullered zone traces of a pattern-welded design are visible on the surface, best seen in places where corrosion has had a more severe etching effect.

The pattern-welding runs the full length of the fuller and consists of three parallel composite rods welded side by side, together occupying the full width of the fuller. Each composite rod was alternately twisted, then left straight for intervals of about 30mm, and then welded side by side so that the straight and twisted portions alternated across the width of the fuller. An x-radiograph indicates that two triple sets of composite rods occupy the surface, one set on each side of the blade. A close inspection of the more deeply corroded parts of the blade where the pattern-welding is visible,

The outstandingly imposing Gilling West Saxon (English) sword, notable for its revolutionary design and entrancing Trewhiddle decoration. Note the rounded, acutely curved, undecorated guard. About AD 866. (York Museums Trust)

especially where a few small bits are missing, appears to show that each composite rod occupied half the thickness of the blade. This indicates that the two triple sets of composite rods were welded back-to-back and not to a separate central core piece, as is often the case.[1]

Figure 2A (overleaf) shows a schematic impression of how the pattern-welded design would have appeared on the surface along the upper and central parts of the blade, and Figure 2B gives a three-dimensional view across the blade. The twisted parts of the pattern-welding show both as a diagonal 'grain' and as a more distorted 'watery' pattern, The 'watery' pattern occurs mostly nearer the hilt end of the blade and shows that here the fuller was ground away after the final forging. The reconstruction of the design shows a straightforward diagonal grain as the alternative to the 'watery' pattern, and although this appears to be largely the case the diagonal graining does appear to be rather distorted in places, tending to resemble the 'watery' pattern. This indicates that lower down the blade, in the fullered zone, some of the surface was ground away after final forging, but less than nearer the hilt where the effect is quite pronounced. Towards the tip the diagonal graining appears largely undistorted indicating that little or no surface grinding took place.

These different aspects of the pattern-welding are visible in the areas of hard, shiny black patina

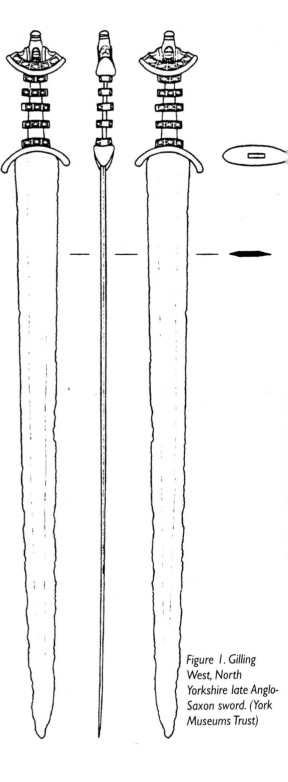

Figure 1. Gilling West, North Yorkshire late Anglo-Saxon sword. (York Museums Trust)

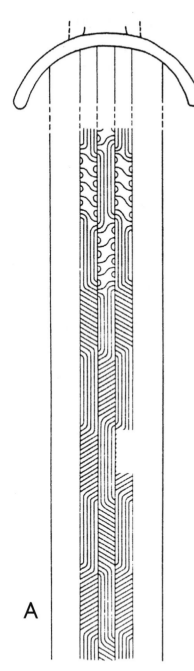

which preserves the original surface of the blade. This is a fairly clear indication that the blade was heavily etched after any final surface grinding and polishing, so that the pattern-welding design would have been visible. The etching has left alternate bands in the pattern-welding as higher and lower ridges, which indicates that the iron of these bands contained differing alloying elements that probably alternated fairly evenly in composition. The black, shiny patina, however, only preserves the shape of the surface and not its original appearance or colour. Along the pattern-welded area this would have shown as alternating paler and darker bands, further accentuating the already alternating nature of the design. The cutting edges of the sword may have appeared as paler or darker zones after this final etching depending on the etchant used, the alloying elements present in the iron (mainly the proportions of carbon or phosphorus), and any final heat treatments to which the blade may have been subjected. A number of cracks running at right angles across the cutting edge along one side near the hilt may suggest that the blade was heat-treated and that the cutting edges either wholly or partially consist of a steel with a carbon content high enough to become much harder and more brittle on quenching.

The well-preserved nature of the Gilling West sword meant that no sectioning for metallographic analysis was permissible so it is difficult to say much more about the structure of the blade than the observations given here. It is not possible to say to what extent steel was combined with wrought iron in the manufacture of this weapon or to evaluate the quality of the metal used to fashion

Figure 2. Gilling West Late Anglo-Saxon sword. Schematic views of the blade to show the probable structure of the pattern-welding on the upper and central part of the blade (A) and in section (B). Based on surviving surface detail and X-radiographs. (York Museums Trust)

the blade. Metallographic analysis on sword blades of a similar date and type would suggest that this blade is very likely to have included welded-on cutting edges of heat-treated (i.e., quenched and possibly tempered) steel.[2] The pattern-welded zone along the fuller on either side must have given a highly decorative appearance to the blade and one which was clearly intended to be seen.

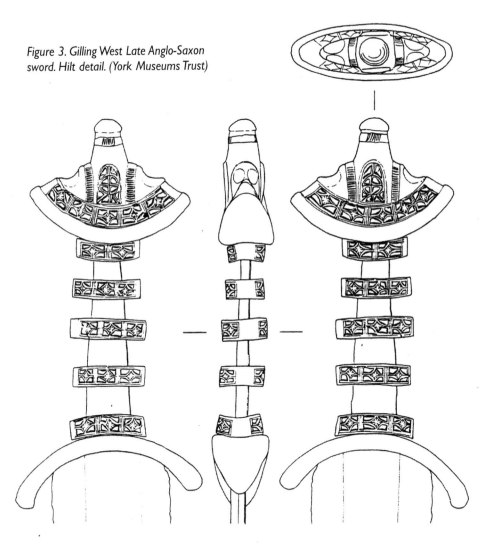

Figure 3. Gilling West Late Anglo-Saxon sword. Hilt detail. (York Museums Trust)

Chapter Eleven

Armour Protection and Cavalry Manoeuvres

This chapter examines the form and efficiency of armour protection and horse and cavalry manoeuvres of the English and Norman French in the second half of the 11th century, drawing largely on information and deductions gleaned from the Bayeux Tapestry. We begin with the armour of the English aristocracy and thegns and the body protection of the men of the Saxon fyrd and general levy. Then we look at the armour of the mounted troopers of the Norman French and of their foot-soldiers and archers. I hope that this section will correct some misunderstandings and misconceptions of these subjects. Finally, we shall consider the differing ways in which the English and Normans employed their horsemen.

ARMOUR AND BODY PROTECTION

For centuries western European (continental) rulers were concerned with the provision of well-armed and well-horsed soldiers (milites) from the ranks of their lesser nobility or gentry. 'The expensive equipment of such elite warriors was common to all these societies – a horse with saddle (and stirrups), a bridle, a shield, and a light "lance" (javelin), a sword, a helmet and byrnie or hauberk.'[1] Probably, the most significant widespread introduction of this process was instituted by Charlemagne, later Emperor of the Franks, during the 9th century. He stipulated in his Capitulare missorum of 792–3 that the accoutrements of every cavalry vassus (mounted soldier) were to be standardized as a horse, lance, shield, sword and hauberk (mail coat). In the 9th and 10th centuries these had become the essential equipment of mounted Frankish troops and those of other western countries, enabling them to protect western Europe from the Avars and Vikings and to achieve victories over the Saxons, Moors and Arabs.

The Anglo-Saxon kings of England and the Norman dukes were anxious to ensure that as many of their warriors as possible were well armed and protected.

'Another indication that the Anglo-Saxon nobility like that of Normandy and the rest of northern France was essentially a military nobility may be found in the fact that under English law every man of thegnly rank or higher was obliged to arrange for the payment on his death of his military equipment to his lord.'[2]

'The payment known in English as heregeatu, heriot of war gear, comprised for the ordinary or lesser thegn his horse and his weapons.'[3]

Norman French cavalry advance: two hold their spears in overarm throwing position. Archers follow behind, only one of whom is armoured. (Detail from the Bayeux Tapestry – 11th Century. Reproduced by special permission of the City of Bayeux)

This legal obligation against a warrior on his death ensured the maximum amount of armour, weapons and related war accoutrements were distributed to other warriors who previously lacked them. Readers will appreciate that there is some similarity between this system and that prevailing during the Germanic Migration period, when lords gave swords to their war band followers. 'From the evidence of the secular code of King AEthelred's successor, Cnut, it appears that the heriot of the lesser or median thegn comprised one horse and its tack, one helmet, one byrnie or mail coat, one sword and one shield.'4 'A concern for the supply of helmets and byrnies (mail coats) as of ships is also indicated in the will of Archbishop AElfric (1003/1004) which bequeaths to AEthelred his best ship and sixty helmets and byrnies.'5 The Anglo-Saxon Chronicle of 1008 states that the King gave orders that 'every eight hides' were to provide a helmet or a corslet. The most interesting fact emerging from surviving manuscripts on this subject is that an English thegn was a very-well-equipped mounted warrior by the late 10th century. Furthermore, his equipment, with the exception of the Anglo-Scandinavian two-handed battleaxe, was almost identical to that of his Norman French military counterparts.

'The weapons of the late Saxon society in England denote a wealthy aristocracy capable of exercising extensive aristocratic patronage. They also show that an English army had been developed whose armour and weapons were comparable with that of the milites (well armed, carefully selected warriors) of the feudal world of northern France.'[6]

THE BAYEUX TAPESTRY

This provides military historians with considerable and very useful data about the pre-events to the Battle of Hastings, the course of the conflict, and the weapons and the accoutrements employed by soldiers on both sides. It is the best contemporary illustration of protective armour. It is not actually a tapestry, but an embroidery, made by English ladies and maids in Kent. The tapestry, which was made on the order of Bishop Odo, measures 231.5 feet long (70.34 metres) and 19.5 inches wide (50 centimetres). It is stitched in eight colours on coarse linen. It was hung for a long time in Bayeux Cathedral. In his paper 'Arms Status and Warfare in the Late Anglo-Saxon England' N P Brooks makes two particularly interesting references to this:

> 'We find that the English soldiers are shown through the battle as well-armed warriors with which we have seen to characterise the nobility – the thegns. They wear conical helmets and trousered byrnies; their shields are predominantly kite shaped, but occasionally round or sub-rectangular; they have spears and swords and sometimes two-handed battle axes.'

> 'But it is at least clear that the tapestry artist distinguishes between the main body of the English army, which formed the shield wall and comprised soldiers with a full complement of the weapons and body armour, and the shield man who played a subsidiary role. The artist makes the same distinction between the noble and non-noble soldiers in depicting the Norman army with its uniformly well armed knights and its archers who with single exception have no body armour at all.'

Armour

Norman and French horsemen wore a knee-length, one-piece mail shirt, with three-quarter-length sleeves, over a padded hauberk, which also protected the neck and throat. The bottom mail section was like an open skirt, as a mounted trooper did not need to protect his inner leg because it was not exposed when he was in the saddle. 'Furthermore, the discomfort of riding in mail trousers might be palliated by padding, but damage to saddle and horse from friction could not.'[7] They protected their heads with an iron helmet fitted with a nasal guard. The English housecarls used almost identical armour. They also wore a protective garment of interlinking ring mail and a conical iron helmet with nasal guard. The main difference was that when the English fought on foot they needed to protect their groin area and thus laced up their lower mail coat to make trousers. By the 11th century mail was made with either round or flat rings riveted together to make a suit. Perhaps the strongest defence was achieved by a mail incorporating a mix of both flat and rounded rings. The weight of a mail coat stretching to the waist was

This clearly shows the similarity of housecarl and Norman/French armour and shields. On the left an English unarmoured archer is dwarfed between two housecarls who carry behind their shields three spare javelins. These were particularly devastating in halting Norman/French attacks against the shield wall. (Detail from the Bayeux Tapestry – 11th Century. Reproduced by special permission of the City of Bayeux)

about 28 to 42 pounds; a full-length coat weighed considerably more. Below the mail they wore a leather coat or padded jacket, which provided an essential additional safeguard. These protected a man from the contusion effects of a swinging sword or a mace blow, and much reduced the effects of a spear thrust or javelin strike. However, mail would not halt arrows. 'A mail coat would be most effective if worn over a leather jerkin, though evidence for such an undergarment is absent until the Middle Ages.'[8] Certainly the Vikings adopted this custom from an earlier period.

The armour used by the English and Norman French military elite was therefore very similar. Generally it would have provided some protection against edged weapons, particularly those wielded in a slashing manner, but it would have been less effective against sword- or spear-points. It would not halt an arrow unless it was a long-range one. The mail coat, unless worn over a quilted jacket, would be vulnerable to a firm thrust from a very long spearhead of the type used by the English, which was designed to penetrate armour. Doubtless these inflicted casualties at Hastings. However, the large and very robust kite-shaped shields would safely halt short-range arrows, sword cuts and spear thrusts. This is clearly confirmed by the Bayeux Tapestry, which shows housecarl shields sustaining numerous hostile arrows. These shields, often made from lime-wood and covered with leather, had a maximum width of about 15.5 inches. They provided very good personal protection.

With one exception, depicted in panel 60 of the Bayeux Tapestry, no Norman French archers are shown wearing armour. Their apparent lack of this resulted in high casualties at Hastings, as we shall see in Chapter 12. Their infantry, who are little depicted in the tapestry, carried swords and spears, wore mail and perhaps helmets, but despite this protection also suffered at Hastings. William of Poitiers stated: 'The attackers (the Norman French) suffered severely because of "the easy passage" of the defenders' weapons through their shields and armour.' This indicates their poor and ineffective quality. The English militia also generally had no armour except a buff coat, although its extent depended on the wealth of an individual. They did possess the useful round shield, with its projecting iron boss to protect the left hand. Because this only screened a small area of the body it would have been necessary to move it promptly to block a sudden blow from an unexpected direction. The unarmoured English militia were susceptible to short and high-angled arrow fire and to mounted troopers if isolated in the open. However, when fighting in the shield-wall formation they were more secure and, with their spears of various lengths creating an intimidating frontal mass of gleaming blades, presented a formidable defence to any attacker. Furthermore, men of the militia were often adept at agile ducking-and-weaving techniques, and would sometimes, at close quarters, avoid a hostile blow then at once deliver a successful one with spear or long chopping scramasax.

The Norman French brought a corps of some two to three thousand professional archers to England carrying short bows with a range of about 150 yards. This weapon had little resemblance, in form or use, to the later-famous longbow used so successfully by Welsh and English archers. The English army had only a few bowmen, in the normal Anglo-Saxon military tradition, because the weapon was not highly regarded. A possible explanation for this may come from David Howarth, who wrote:

> 'Bows in England were aristocratic sporting weapons, used for shooting deer, not human enemies, and archery was a strictly guarded mystique. For a long time past, the nobles had been fanatically jealous of their hunting rights, and if a poor man owned a bow he labelled himself a poacher. No common soldier in the fyrd would be any good at archery, or admit it if he was.'[9]

It is thus ironic that at both the Stamford Bridge and Hastings battles the bow played a crucial part: King Harald Hardrada of Norway was fatally struck in the throat by an arrow, probably fired by a Yorkshire huntsman; King Harold II of England was later, possibly, wounded at the battle's crisis point by another at Hastings.

Cavalry

The Norman French cavalry comprised troopers, who rode fairly sturdy but unprotected horses called 'Destriers'. These were bred and drilled to carry an armoured horseman. Such troopers were called 'Miles' or 'Chevaliers'. In England, military aristocrats (thegns) riding horses were referred to as 'knights'. Clearly, mounted troopers of the 11th century, with their light lances or throwing spears

Housecarl axemen combating cavalry. Three housecarls are correctly shown carrying swords, with which the vast majority of these elite troops were equipped. Axemen usually operated in pairs with one man dealing with the trooper's weapon, which made them very effective. (Detail from the Bayeux Tapestry — 11th Century. Reproduced by special permission of the City of Bayeux)

bore very little resemblance to the true armoured knights of the 13th and 14th centuries, being neither equipped nor trained to achieve shock power. Eventually cavalrymen of the 14th century often wore intricately fashioned suits of plate armour, and closed helmets that afforded considerable protection. Their armoured steeds, such as the percheron or Clydesdale, were large and powerful, capable of carrying the weight of armour while reaching sufficient speed to make effective charges. These extremely expensive animals were also carefully bred and trained. The true knights used a strong couched lance and were trained to charge together, knee-to-knee in an irresistible and unwavering formation. 'The horse-soldier of the eleventh century had not reached that advanced stage of military development.'[10] The Norman French cavalry were thus necessarily more circumspect and cautious owing to their inadequate armour and unprotected steeds. Doubtless, Norman French horsemen were periodically effective in battle against other cavalry because they were trained to fight them on a one-to-one basis. They might also deal effectively with isolated small groups of unprotected peasants equipped with shields and spears. However, against well-trained, experienced and professional infantry on the defensive and forming a firm shield wall, they were certainly not.

The Bayeux Tapestry depicts troopers advancing on the English in somewhat scattered groups called 'contoi' carrying differing weapons such as maces, swords, light spears or javelins. To be more effective they should have all, perhaps, carried spears and advanced in larger, more compact groups.

On approaching the enemy, the drill was to launch light lances from some distance in the hope of inflicting casualties before turning away, allowing comrades to repeat the manoeuvre. Their purpose was to create a break in the defence line into which they would advance, and attempt to widen it with slashing swords. However, we shall see in Chapter 12 that until the shield wall eventually disintegrated, most horses approaching the English line too closely might have been decapitated and their rider killed with an axe or spear. Troopers may have attempted a forward spear thrust but most horses would naturally have shied away from the bristling and dense line of spears. It is not surprising that many Norman French warhorses could not withstand the heavy and continuous fusillades of missiles that greeted attacks some 20 to 30 yards from the English shield wall. So the Norman French cavalry did not at all resemble the later, fully developed feudal heavy cavalry. However, we should remember that the steep and boggy ground at Hastings was disadvantageous to horsemen.

It is interesting that the English cavalry used their horses in a very different manner from that practised on the Continent. The principle in England was that warriors rode to battle, fought dismounted, and then rode home after the conflict. From the time of King Alfred, the primary purpose of Anglo-Saxon cavalry

A scene from the Battle of Brunanburh (AD 927), where King AEthelstan of England defeated a huge hostile confederation army led by Constantine, King of Scotland. 'A Norse Jarl is mortally wounded. His hearth troops attempt to spirit him away as West Saxon warriors move in for the kill.' (From an original painting by the late Rick Collins; reproduced by the courtesy of Military Modelling magazine)

(mounted infantry) was to move a small, mounted infantry army to an area of conflict at sufficiently high speed to confuse and surprise their adversary. They then fought on foot. Harold achieved this on his Welsh campaigns. The tradition of mounted Anglo-Saxon armies moving at great speed to confront an adversary continued throughout their history and the tactic was usually very successful. However, the procedure had one snag: the dismounted fyrd troops would naturally be unable to keep up with the mounted ones, which consequently reduced the size of the force that eventually fought the battle.

Another important role of the Anglo-Saxon cavalry was to provide a small, mounted mobile reserve to be employed only when an adversary started to display weakness in a battle that was being fought on foot. These troops then used their horses to convert a retreat into a rout. A classic example of this was at the Battle of Brunanburh, in 937, when King AEthelstan defeated the huge northern confederation host. According to the Anglo-Saxon Chronicle, the English promptly followed up their victory (fought on foot) with a long pursuit by mounted infantry:

> 'All through the day the West Saxons in troops
> Pressed on in pursuit of the hostile peoples,
> Fiercely, with swords sharpened on grindstone,
> They cut down the fugitives as they fled'

> Anglo-Saxon Chronicle, AD 937; p.108

It is often assumed that, because the English fought on foot at Hastings, they therefore always fought on foot, and employed horses only to reach, and then later depart from, a battlefield. At Hastings, of course, Harold wisely dismounted his housecarls in order to create a very strong defensive line. 'Much as he would have liked, no doubt, to have kept his mounted housecarls in reserve, poised to launch a counter attack, the rather poor quality of his rustic army demanded that he dismounted his professional soldiers to stiffen the ranks of the militia.'[11]

We conclude this chapter with mention of swords used by the housecarls and Norman French troopers. Those employed by both sides were similar. They were often of the double-edged type with fullered blades similar to those of Wheeler's Type VII. English ones possibly retained more Viking features than those of their opponents. Some swords had imported inscribed Rhineland or high-quality English blades and were fitted with short, straight guards and beehive or Brazil nut pommels. Some patterns may have incorporated thick, faceted 'wheel' pommels, sometimes with bevelled edges, or the three-lobed, cocked-hat type. Some swords would have had a much longer guard, in a variety of simple but slightly differing forms, similar to those of Oakeshott's Type IX. These were designed to provide better hand protection from opponents' sword cuts. This type was to be a characteristic of the subsequent knightly swords. Swords of the late 11th and early 12th centuries tended to have rather larger pommels designed to act as counterweights to the rather longer, more tapering blades.

Chapter Twelve

The Hastings Campaign

On 25 September 1066, King Harold of England achieved a decisive victory at Stamford Bridge near York, in a battle that is significant because it represents the culmination of the battle tactics and weaponry of the period and, as such, warrants some study here. He defeated King Harald Hardrada of Norway, the famous military veteran commander of a large, battle-experienced Viking army. By means of driving and charismatic leadership Harold galvanized his large army, which included part-time soldiers, into achieving an astonishing forced march from London to York. The speed of these well-co-ordinated movements enabled Harold, with his entire army, to confront and completely surprise the Norwegian king at Stamford Bridge. Hardrada was unprepared and had only about two-thirds of his host on the field. These advantages directly contributed to the subsequent stunning victory. However, the battle was long, hard and bloody, and the English losses, particularly of less-well-armed soldiers of the fyrd, many of whom lacked battle experience, were very high. The English regular army comprising the military elite housecarls, 'those professional soldiers who represented the maximum of military efficiency to be found in the Anglo-Saxon Danish world',[1] may also have sustained significant casualties. The captured Viking fleet of some 250 ships, along with Hardrada's petty cash – a gold bar that needed several men to carry it – were small consolations.

On 1 October, whilst celebrating with the mandatory victory feast at York, Harold received news that William of Normandy had landed at Pevensey, on the south coast, on 28 September. After some reorganization in York, he returned rapidly to London with the housecarls and other mounted warriors. Messengers delivered fresh call-up instructions for military reserves ordering troops in the Midlands, the western counties and the southern counties north-west and east of London to join him in the capital. The remains of the unmounted southern fyrd straggled behind him. The northern earls, Edwin and Morcar, agreed to follow the army when their much-depleted followers, who had fought at both Fulford (a major English defeat on 20 September) and Stamford Bridge, had been reorganized. King Harold and the regular army reached London on 6 October. Soon, fyrdsmen of the Midlands, western counties and those of the south-east not involved at Stamford Bridge streamed into the city. According to Wace's chronicle 'The men of England flocked to Harold's standard.' So, with time, Harold would assemble another very large host because the nation possessed many available fighting men who had not yet been involved in the national conflict.

Meanwhile, William's well-trained army of approximately 2000 mounted troopers, 3000 armoured infantry and 1000 archers were camped near Hastings.

Their headquarters was a wooden castle protected by ditch and palisade. Being anxious to make battle before the English had assembled a large force, William deliberately undertook a programme of plundering and murder in the Hastings area: a form of 'military' activity at which continental troops excelled. William hoped this would tempt Harold to rush south quickly to the protection of the subjects within his own earldom of Wessex.

ENGLISH OPTIONS

King Harold had the following options for dealing with the Norman threat:

1. To employ the navy to cut the invaders' lines of communication with the Continent to prevent their reinforcement and impede William from re-embarking his army to move to a more advantageous location. During this period Harold should remain on the defensive in London, rest the housecarls and gradually assemble an army of about 16,000 men. Meanwhile, he would make thorough reconnaissance of William's army to determine its size and, in particular, the number of warhorses.

Then, when all his battle preparations had been made, he would advance southwards and give battle on ground of his own choosing with a considerable numerical superiority.

2. To starve out the enemy by moving the English population, animals and provisions from a wide area surrounding Hastings. After allowing time for this to affect the Norman troops, he would advance on William and give battle.

3. To attack at once with the existing, rather small army in the attempt to catch the Normans by surprise, as had been effectively achieved against Hardrada at Stamford Bridge. There was, of course, one major difference between the two situations: at Stamford Bridge the English possessed a large army, which, to a degree at least, increased the possibility of victory should surprise not be achieved.

4. Magnus Magnusson, in his book *Vikings*, considered Harold's surprise strategy theoretically impeccable. 'He knew that Pevensey (Hastings) was a geological peninsula, hemmed in by a ridge through which there was only one exit; he wanted to stop a Norman break-out and keep William bottled up for the winter.' To achieve such a blockade would necessitate a rapid move south before William left the location and advanced north.

Harold's brothers Gyrth and Leofwine offered shrewd and sensible advice. They reminded the King he was exhausted from his exertions in the north and wisely suggested delaying a new campaign until the army had first rested and been reinforced. The tactic they suggested thereafter was a mix of options 1 and 2. Harold refused to be persuaded, however, favouring option 3. He underestimated the size of William's army, particularly of its cavalry, failing to believe that many warhorses could have been shipped across the channel because such an operation was difficult and hazardous. Harold was the fleet commander and well versed in nautical matters and so his judgement on ships was naturally credible. In this case, however, he failed to contemplate construction of unconventional ships as the means of transporting horses. The Normans had built ships powered by sails, and not by oarsmen, which provided more space to carry animals. However, the ships had the serious disadvantages of being immobile without wind, and very difficult, if not impossible, to control in storm winds. To utilize such vessels William took a very great risk indeed. Without wind they could not move anywhere and could only progress if the wind was in the right direction. It is, perhaps, not surprising that Harold failed to visualize construction of utilitarian ships. The King was also anxious to relieve the English population from the effects of Norman plundering. He therefore declined the scorched-earth policy, stating:

> 'Never will I harm an English village or an English home, nor will I harm the lands or goods of any Englishman. How can I do hurt to the folk who are put under me to govern? How can I plunder and harass those I would fain to see thrive under my rule?'[2]

Harold's wisest policy at this juncture and the one recommended by the thoughtful and wary Gyrth was,

> 'to have maintained his position in the capital, and collected there all his levies, while the force already assembled be interposed between London and the invader. Every day's delay would have been a loss to William and a gain to Harold; for with every day's delay the defence would have grown more formidable and the attack less strong.'[3]

Harold also ignored this advice but at least 're-mobilised the fleet and sent it to patrol the channel.'[4]

In the event, the King determined to make battle as soon as possible. He assumed, with some justification, that William would not anticipate an assault so soon after the exhausted army returned to London. One is compelled to conclude that his primary reason was to replicate the surprise which had contributed so markedly to his victory over Hardrada at Stamford Bridge. In this he was, unfortunately for England, over-confident, because the two situations were different, particularly regarding the opposition commanders. William was much more calculating and shrewd than the bold and daring Hardrada, while possessing a definite knowledge of Harold's personality.[4]

> 'Brave and courageous as he undoubtedly was, Harold was too impetuous to be a great military commander. Two weeks before, in September, he had fought a great pitched battle in which his army must have suffered severely. He won it by an impulsive dash to the north, by a march which took the enemy by surprise.'[5]

In the new situation there was no essential element of surprise because William anticipated that Harold would probably attack him as quickly as possible. There was a final factor: the King possessed in good measure the commendable English qualities of dependability and reliability. Therefore, in his appreciation of the campaign he considered the condition of his people to be as important as the vital and far-reaching military concerns. This was a serious emotional mistake.

THE SIZE OF THE ENGLISH ARMY

His courageous and very impetuous decision made, the King left London with the housecarls and fyrd on 12 October. The fyrd were mostly on foot. Consequently, many enthusiastic warriors, still making their way to the capital to join him, were left behind. The King was very sensible that some of the bravest men in England had fallen in the battles of Fulford and Stamford Bridge, and that half of his troops were not yet assembled. The army's strength is not precisely known but has been roughly calculated from the positions it might have subsequently adopted on the Hastings battlefield. 'These were thought to have been about 6,300 men on a 600 yard front (if deployed in 10 ranks deep), or about 7,500 men (if deployed in 12 ranks).'[6] Edwin and Morcar's contingents had not arrived in London in time to join the army. They actually appeared on the road to the city some days after the Hastings battle, when they received news of the conflict. Calculations of English manpower at Hastings are, to a degree, complicated by the fact that English units were evidently still arriving, probably from different directions, at the assembly area the day after the battle had started. Presumably, these were not included in the 'ground' calculations. We shall return to this subject later in the battle section. Harold's host certainly contained many fresh and competent warriors who had not served in the Stamford Bridge campaign and was thus an army of high general standard. The King had selected as the army rendezvous the place of the 'hoary apple-tree'[7], on a rise in the Downs. The town of Battle now stands on this site. He probably reached the point on the evening of Friday, 13 October. Numerous contingents arrived, bone-weary, throughout the night. Indeed, as I have mentioned, several were still reporting the following day. If Harold's purpose was still to surprise William, his army may have spent the night concealed in the woods and only emerged onto the 'hoary apple-tree' location early after sunrise the next day. However, Norman scouts detected the arrival of the army and reported it to William. So there was no surprise, but William was given plenty of time to prepare for his attack.

KING HAROLD'S SURPRISE PLAN

It is important to clarify King Harold's 'surprise attack' plan in the Hastings campaign. Did he actually consider this to be one of his options? We shall draw relevant plausible deductions from the known facts. In this I referred to J F C Fuller's account of these events in his book *Decisive Battles of the Western World*. Fuller is known for his brilliant military battle-appreciations and perceptive deductions drawn from them. We appreciate that the King was over-anxious to save his Wessex subjects from Norman depredations, and underestimated the size of William's force, particularly the number of warhorses that might have been transported across the channel. His incorrect appreciation of the latter strongly indicates that he believed he could defeat William with only a small but combat-experienced army. We can also deduce from his refusal to heed his brother's advice that he was determined to act very promptly against his foe.

'That surprise based on speed was in his mind (Harold's) is supported by his previous generalship. In his Welsh campaigns of 1063 and his Stamford Bridge campaign he had moved like lightning in order to surprise; therefore, probably Mr Round[8] [the 19th century historian] is right in surmising that this was also his intention in the present campaign.'[9]

The assembly area that the King selected was coincidentally adjacent to an excellent spot on which to fight a defensive battle. Its characteristics are examined below. The King probably made a detailed reconnaissance of this and, doubtless, several other locations during the summer while waiting for the arrival of William's fleet.

Additionally, 'It was a focal point of roads, which made it an ideal assembly-area for an army converging from several directions.'[11] Contingents from various areas joining the army would thus have travelled on these. The King could have contemplated making a surprise night attack, launched from this assembly area, on the invaders' camp some six miles away at Hastings. In the event this plan was not feasible 'because he had not reached the Battle area as early as he hoped. To have pushed on the remaining six miles to Hastings after dark, and then attempted a night attack, unpreceded by a daylight reconnaissance, would have been madness.'[12] 'In the circumstances, the only chance to surprise William was to attack at dawn (this would have occurred next day about 5.30 a.m.) which would have demanded an advance soon after midnight. Even if considered, the weariness of Harold's men must have prohibited this.'[13] So we can conclude that Harold probably intended to launch a surprise attack but on account of the factors described was prevented from achieving this. Paradoxically it was to be William who surprised Harold. Just before the conflict he was firstly astounded to see the large number of Norman cavalry troopers and was then caught unawares by the speed at which William launched his first attack. 'William came against him by surprise before his army was drawn up in battle array.'[14] We do not know if Harold intended to fortify his position with sharpened stakes and ditches along the edge of the ridge but, if so, there was no time to do so. Such procedure was not, perhaps, contemporarily usual and so the inability to take such precautions may not have discomforted the King.

THE GROUND

Had Harold always determined to fight a battle on the ridge, bearing in mind his primary objective of mounting a surprise attack? In the event, he did fight there and could be commended for selecting such an excellent position. Fuller makes further revealing comments on this point:

'What was Harold's plan? Was it to surprise William, as has already been discussed, or was it to assume a passive defensive – that is, block the London road and await attack? Simply to hold the "hoary apple-tree" position was not sufficient to rid England of her invaders. Anyway how could the King know that William would attack him there?'

In the event, of course, Harold was compelled to fight a defensive battle for two reasons: firstly, he reached the location too late the previous evening to implement his surprise-attack plan, and secondly, William very much wished to make battle with him as quickly as possible, and moved fast to achieve this. Consequently, Harold had no option but to fight where his army stood. It is sometimes stated that William may have tempted Harold to fight at a time of his own choosing but was compelled to fight on ground of his opponent's choosing. This is a true appreciation of events. But would William have (actually) made battle if he had realized just how very strong the position was and how capable the English would be in defending it? Perhaps he can be considered impulsive for launching a full all-arms attack without bothering to make a thorough ground reconnaissance.

The ground comprised a projecting spur emerging from a large wood. The sides of this narrow spur, which fall precipitously, join an axe-head formed ridge jutting into a boggy and marshy valley. The ridge is about 800 to 1000 yards wide at its highest level.

'The spur and its cross ridge were ideal positions for assembly and defence, for an army drawn up on the cross ridge could not be outflanked or by-passed. It could be assaulted only from the front and sides, where it was protected by slopes of varying steepness. The slopes of the ridge made assault by horsemen hard, and on the east virtually impossible, for there the fall at the back of the ridge is exceedingly steep, almost precipitous, the gradient in the rear being 1 in 4, and at the side 1 in 12.'[15]

It was thus an excellent site for infantry to confront cavalry. The highest point of the site is what is now known as Abbey House, at the front of which a slope drops some hundred feet in about 400 yards to the head of Asten brook, now a series of fish ponds. To the right ridge-centre was a small hillock, some 20 feet above the surrounding ground, giving an excellent view of the valley. On ground that strongly favoured infantry, the English housecarls were obviously the best troops on the field.

ENGLISH ARMY DEPLOYMENT

Harold drew up his army across the ridge peak in the traditional English shield-wall formation. His aim was to maintain a very firm, almost impregnable front line throughout its length that could not be turned on either flank. This would naturally have appeared as a formidable shield-wall array. 'It was an admirable formation against infantry armed with sword, spear, and axe, and an essential one for infantry against cavalry relying on shock.'[16]

Having studied the ground I agree that the protection of the flanks was achieved as follows:

'It is highly probable that he (the King) occupied the 600 yards between the little brook west of Abbey church and the junctions of the Hastings and Sedlescombe roads in order that his flanks might rest upon the two steep depressions.'[17]

Fuller also suggests that Harold's standards were originally placed on the ridge of the summit 'which at the close of the battle fell at a spot seventy yards to the east of the Abbey House, and was marked later by the high altar of the Abbey church.' Regarding the manpower strength and its deployment, we can reaffirm that: 'If Harold drew up his army in a phalanx of ten ranks deep, allowing two-foot frontage for each man in the first rank – the shield wall – and three-foot frontage for those in the nine rear ranks, then, on a 600 yard front his total strength would be 6300 men and, if in twelve ranks, 7500.'[18]

We must remember that additional English troops arrived after the conflict had started, so perhaps the army was actually rather larger. Having 'fought' the Hastings battle several times in the past, I consider the English strength would have been in excess of 7500. The front rank may possibly have been fairly heavily manned by housecarls. However, it has also been suggested that these elite soldiers held the second rank and emerged only when the enemy had advanced to close quarters. Among these would have been the feared axemen, who would suddenly have formed an intimidating front line. Axes required considerable space to wield effectively and achieve the severing of infantry heads, troopers' left legs, and the dismembering of horses, producing spectacular fountains of blood deliberately calculated to lower Norman morale and enthusiasm for the fray. Behind the massed ranks deployed housecarl hammer-throwers who launched huge stones at up to about 50 yards. Their effect was disconcerting and unsettling as they descended from the zenith and fatally struck a target; they could also kill a trooper and horse simultaneously.

English horses and ponies were tethered well to the rear of the spur near the woods. Doubtless contingents arriving after the battle started were met at this

The shield-wall, fronted by housecarls, stands firm whilst delivering a constant fusillade of assorted missiles. Norman French cavalry and infantry attacks were incapable of breaking the English defence until much later in the conflict. (Detail from the Bayeux Tapestry – 11th Century. Reproduced by special permission of the City of Bayeux)

point and led forward to the front line when appropriate. At this battle the English did not retain a mobile cavalry reserve because all the best warriors were needed to maintain an impregnable shield wall and stiffen morale of less-experienced troops.

THE BATTLE

Before the battle started, Harold rode down the front rank of his army issuing strict instructions that no man was to advance forward from the shield wall without orders. Should they do this, he said, they would be isolated and then probably slain by a Norman horseman.

It is usually accepted that the battle started at about 9 a.m. If this is correct, then, as Fuller perceptively comments, William must have set off early. After reckoning the time taken for assembly, march and deployment, he must have left Hastings at between 4.30 a.m. and 5 a.m. However, as the sun did not rise until about 6.20 a.m. on that morning, it seems possible that the timings were actually later. Therefore, the conflict may have started at about 10 a.m. If so, it is unlikely that Harold was caught on the hop, and the Anglo-Saxon Chronicle's comments on this battle stage are wrong. Did Harold take advantage of the time to strengthen his position?

Phase I: The First Norman French Attack

The Norman army's order of battle was: on the left flank the Bretons, commanded by Count Alan of Brittany, on the right flank the French and Frankish mercenaries, commanded by Eustace of Boulogne, and in the centre the Normans commanded by William. According to Fuller, the latter did not actually lead his troops, no doubt wisely, but advanced in the safety of the rear of his formation. The army deployed archers in the front ranks, followed by the so-called heavy infantry and, finally, the cavalry.

On approaching, archers loosed their missiles but, because they were firing uphill, these either flew harmlessly over the English ranks or were taken by the robust shields. 'The English resisted valiantly and met the assault with showers of spears and javelins and weapons of all kinds together with axes and stones fastened to pieces of wood.'[19] Doubtless these periodically included the large, lethal stone hammers. 'The shouts both of the Normans and the barbarians (English) were drowned in the clash of arms and by the cries of the dying, and for a long time the battle raged with the utmost fury.'[20] Naturally the English took advantage of the higher ground. They presented an impregnable front, and their use of enormous axes, aimed at the Normans' unprotected sides, helped them to dominate the battle and convincingly deal with the Norman heavy infantry. The seven-foot axe sweep would have out-reached the Normans' weapons, and the Norman warriors would have been easier adversaries than the towering Vikings at Stamford Bridge with their notions of reaching Valhalla. William of Poitiers further states that the English 'profited by remaining within their position in close order.' He also makes a revealing and interesting comment on poor-quality French body armour, stating: 'The attackers suffered severely because of the easy passage of the defenders'

Re-enactment: a Breton trooper, who has just hurled his spear, confronted by English axemen. (John Eagle)

weapons through their shields and armour.' This confirms the good design and quality of many English spearheads and the high tempering of their swords. The famous axe, of course, would cut through any defensive clothing including their ring mail and helmets. It also, perhaps, indicates the rather poor-quality produce supplied by the continental arms industry. William of Poitiers makes an additional, slightly ambiguous statement: 'The English bravely withstood and successfully repulsed those engaging them at close quarters, and inflicted loss upon the men who were shooting missiles at them from a distance.' This must be correct because, during the English missile fusillades, front-rank French archers would have been slain, particularly as they were unarmoured, but so would the heavy infantry, troopers and horses towards the rear of the attacking formations. It

would be interesting to know why he only specified archers. Perhaps he also referred to light lances thrown by troopers from a distance, which the English would have returned with a vengeance. To summarize the general battle situation at this point, the English had most decisively withstood the first main Norman French attack, which seems to have had little or no effect on the shield wall's stolidity.

We know that the English missile fire was effective. We can assume that, at the start of the battle, the English were well stocked with an assortment of missiles. Presumably, to ensure a dense cloud of effective missiles was created, these were not thrown until the enemy's front rank was only about 10 to 15 yards away. The missiles' range could have been from about 30 to 50 yards. Assuming that an enemy formation attacked a section of the shield wall with a manned frontage of 300 men and only the first four ranks threw missiles, this means they could generate a cloud of about 1200 throwing weapons – a formidable fusillade. The differing missile types and ranges could thus reach the first six to eight enemy ranks and seriously discomfort them. An axe, for instance, would have a shorter range than the lethal javelin, whereas the light throwing spear might have been the longest-ranging weapon. The effect of the dense weapon cloud would be devastating, particularly as there would be little chance in the packed ranks to avoid them. Providing each Englishman retained a supply of throwing weapons, and one frame of the Bayeux Tapestry depicts two housecarls, each holding three spare javelins, there would be sufficient to create several missile clouds. The problem of running out of projectiles was resolved by warriors going forward after an attack, to collect weapons previously used, which was a gruesome task, but provided opportunities to kill the Norman wounded.

Phase II: The Breton–Norman Crisis

The most serious reverse suffered by the invader during the conflict was an interlude well documented by William of Poitiers, who said: 'The foot-soldiers and the Breton knights (on the Norman left wing) panic-stricken by the violence of the assault, broke in flight before the English and soon the whole army of the Duke was in danger of retreat.' Fuller suggests:

'The Norman left wing got into difficulties, and the English right, or part of it, suddenly counter-attacked, and swept the Breton archers and infantry down the slope so they carried away with them in their flight the knights in the rear. The disorganised left-wing retreat caused the Norman division to become panic-stricken because their left flank became exposed. So they likewise retreated, no doubt also anxious to avoid more high casualties. The Norman right-wing division withdrew. This was the battle crisis; the army recoiling down the slope in some disorder.'

Fuller credibly comments on this decisive battle stage:

'Had he (Harold) now seized his chance, he would have ordered a general advance, and pouring down the slope on both sides of the Hastings road would, almost certainly, have annihilated the Norman archers and infantry. True, Norman cavalry would have got away, but bereft of their infantry, in all probability they would not have drawn rein until they found security behind their stockede at Hastings.'

A further favourable event for the English occurred simultaneously when the retreating Norman cavalry unhorsed William. This further unnerved those Normans who saw it. Although the danger of the incident was only momentary, because he managed to mount another horse, it could have been critical if the whole English centre had been advancing down the slope towards him. Indeed, such incidents were often contemporarily sufficient to defeat an army. So, Harold failed to seize an early opportunity to win the battle.

We examine now in more detail the events on Harold's right wing. English fyrdsmen probably burst through the housecarls' front line (assuming they fronted the shield wall in this area) and effectively drove the whole Breton wing to the bottom of the slope. Had they halted at this point, all would have been well. Unfortunately for the English a significant number continued their advance across the valley bottom causing casualties as they went. At the hillock, or small knoll there was an extremely marshy dell (see map) in which the cavalry became bogged down. In the Bayeux Tapestry a written comment above the hillock says: 'Here fell both English and French.' Fyrdsmen pulled troopers from their mounts and killed them with their scramasaxes while others were slain with spears. Some Breton troopers escaped back to the opposite side of the valley where their appearance caused varlet camp followers to panic and flee. At this crisis point, William personally and competently managed to gather his own cavalry. Ordering them to follow, he led them in a charge against the scattered English footmen. Although fyrdsmen may have killed some horsemen the majority were slain by sword or light lance. Ill-disciplined foot-soldiers are no match for cavalry. Few thus managed to return to the safety of the shield wall, from which the housecarls had watched powerless as so many men were fruitlessly lost. Harold, who failed to grasp the fleeting but vital chance to gain an early victory, had instead suffered a significant

The marshy dell where Breton cavalry and, later, English footmen, suffered high casualties. (Detail from the Bayeux Tapestry – 11th Century. (Reproduced by special permission of the City of Bayeux)

reverse. We do not know how many were involved in the extended advance. William of Poitiers mentions 'several thousands', which seems exaggerated, but he then states that, after the loss, 'the English scarcely seemed diminished in number.' There are three possible solutions to this conundrum: perhaps more fyrdsmen than previously thought did manage to reach safety; Harold may have drawn men from his centre and left wing to reinforce his right, or perhaps men who arrived late at the assembly area were deployed there.

I have discussed with colleagues the interesting but unlikely possibility that the right-wing commander (one of the King's brothers) may have ordered an advance which would have included the housecarls. This would have been contrary to Harold's orders, of course, but one English offensive tactic actually did involve the advance of an army's wings before that of the centre. Perhaps the local commander, anticipating victory on his wing, did order an advance, expecting the centre to follow his example. Unfortunately, we have no evidence to clarify this theory.

Phase III: Norman Cavalry Attacks

After the alarms and excitements of Phase II, both sides probably frantically reorganized their forces. The English shield wall remained secure but, possibly, men were withdrawn from some areas to reinforce the right wing. Doubtless many warriors were sent forward to collect missiles. William ordered another lengthy archer fusillade. As before, this achieved very little because most of the arrows passed over the English army or were caught on the shield wall. William withdrew

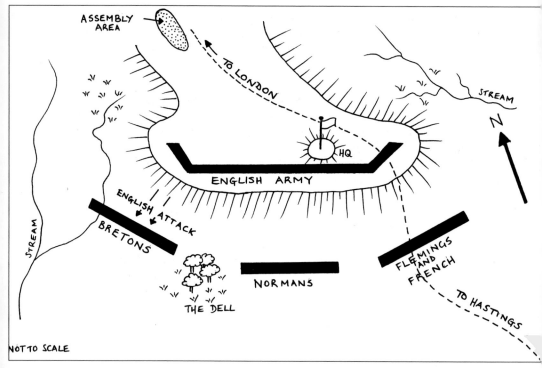

The Battle of Hastings

the remnants of his heavy infantry from the front line. Does this indicate they suffered serious casualties in Phase I? He then ordered a cavalry attack. The troopers, full of confidence after their success against the fyrd footmen near the hillock, advanced up the slope.

They were greeted as before, with a huge roar of verbal English derision, which probably unsettled the horses, followed by a dense missile cloud which inflicted casualties and slowed their momentum. The troopers threw their spears overarm, which sometimes killed a fyrdsman. Because the ranks were so tightly packed, such a casualty was sometimes supported upright by his comrades until the fighting allowed him to be lowered to the ground. It has been suggested that some troopers may have at this point couched their light lances in their attempt to charge a pre-selected Englishman lurking behind the shield wall. This is not true. The prerequisite to such practice was the generation of shock power and this could hardly have been achieved by lightly armoured horsemen on unarmoured horses moving up steep slopes while negotiating piles of their dead comrades and warhorses. The Normans strove to create gaps in the shield wall while the English fought as hard to prevent them. When a gap was created, troopers were confronted and repulsed by the thrusting spears of the shouting fyrd. At close quarters warhorses were easily wounded or killed, and the wounded trooper despatched with a single, firm downward spear thrust. It has been suggested that this battle phase continued for a considerable time without much success for the invaders. It must have been an unpleasant experience for troopers accustomed to riding down continental peasant militias, who, if they survived, always remembered the lesson. 'Never had continental horsemen met such infantry before.'[21]

Phase IV: The Feigned Retreats

The second, smaller Norman battle crisis occurred at about 3 p.m. The English army stood unbroken. The slopes beneath them were littered with the invaders' dead and wounded and the ridge crest was also piled high with decapitated warhorses, troopers, Norman infantrymen and English dead. The English had lost many fyrdsmen from their right wing when they had rashly charged too far forward early in the day. However, the Norman casualties may possibly have been higher at this stage of the battle. William's well-trained troopers were tired and their horses blown. Generally victorious in continental battles, they had proved incapable of defeating defensive infantry despite their lack of archers. I consider it possible that housecarl pairs had worked well together to deal with mounted troopers. They had the advantage of holding the same ground space as one horse; therefore two men confronted only a single trooper. I surmise that the left-hand housecarl would take an advancing trooper's lance, or sword, on his shield whilst his right-hand partner would skip forward swinging his axe in a wide arc to remove a trooper's left leg just above the knee. Sometimes he might remove the horse's head with one blow and kill the trooper with the next. The pair would await the next trooper and repeat the process.

In desperation, William decided to employ different tactics. To understand these better we should consider William of Poitiers' comments: 'Realizing that they (the Normans) could not without severe loss overcome an army massed in close formation, the Normans and their allies feigned flight and simulated retreat, for they recalled that only a short time ago their flight had given them an advantage.' According to Fuller, 'William determined to lure his enemy down from the "Hoary Apple Tree" by means of a feigned retreat.'

The best location for these tactics was probably beneath the English left flank. A cavalry withdrawal from here eventually led to a hill on the Norman side of the battlefield. At this point Norman cavalry would thus be able to turn about and confront their pursuers from a slightly dominant location when the English ran uphill towards them. Another advantage was that cavalry in the central Norman formation could attack the pursuing English on their right flank. William therefore employed this tactic twice and was successful on both occasions. Despite the serious English casualties inflicted, William of Poitiers mentions that the English army 'was still formidable and very difficult to overwhelm.' Nonetheless, the English left wing was much diminished, for which reason Harold re-sited his banners to that area to raise morale. It is easy, perhaps, to blame the English king for failing to retain his troops behind the shield wall on the ridge. However, one must remember that a significant proportion of his brave warriors were inexperienced, part-time soldiers who belonged to the fyrd and thus lacked military knowledge and discipline.

Phase V: The Finale

William realized that, unless he broke the shield wall by dusk, he would lose the battle and subsequently his conquest attempt. Once night fell the English would withdraw under cover of darkness to the safety of the forest. Thereafter, they could move rapidly on tracks known only to them, beyond the range of Norman cavalry patrols. Subsequently, Harold would be able to mobilize another, much larger, army and fight again at a time and place of his own choosing. Meanwhile, the English fleet would prevent both William's evacuation and his reinforcement.

Then William, a master of cunning plans and wily ruses, conceived a new tactic. If his archers shot arrows high in the air above the English ranks so that they fell vertically, their soldiers would be compelled to raise shields and thus be distracted from actually fighting. During these fusillades the cavalry would simultaneously charge, causing the distracted defenders to be less able to maintain the shield wall's stolidity. This is precisely what happened. The archers, leaving gaps in their formations through which the cavalry advanced, ran forward to within 100 yards of the English line and loosed off clouds of high-angled arrows. The cavalry cantered forward under this covering fire.

As the English now occupied a much smaller area than hitherto there was an exposed sector of ground on their left flank. Here, troopers were able to generate a little more momentum in their charges. Many Englishmen, particularly of the unarmoured fyrd, were killed or maimed by arrows. The exhausted foot-soldiers had become rather tightly packed, making it more difficult for the housecarls to swing their swords and axes. Archers were now firing at them at point-blank range. The cavalry made a concerted charge, devastating in its effect, towards the standards. The shield wall at last started to crumble. Four troopers broke through the cordon of Harold's personal guard and struck him down and then, in the true spirit of continental chivalry, hacked him to death on the ground and cut off one of his legs. It has, incidentally, recently been established that the King was probably not struck in the eye by an arrow. This long-held erroneous theory arose because of incorrect interpretation of Bayeux Tapestry panels. With their monarch and his two brothers slain there was nobody to take senior command. The fyrd therefore realized that the battle was lost and many withdrew from the field, some on horseback.

However, a few housecarls continued fighting in a disciplined manner. According to Fuller they 'showed so bold a front that the French commander, Eustace of Boulogne, signalled his men to fall back.' This probably occurred at the time the Normans suffered a serious reverse at a place called the Mallfosse while pursuing the English just after sundown. This seems to have comprised a high, steep bank or causeway, beneath which was a deep, concealed bramble-covered chasm or ravine. In their haste, many troopers galloped to this spot and fell a considerable height from the causeway to the rocks below.

> 'But the growing grass covered an ancient causeway where the charging Normans fell headlong in large numbers, with their horses and arms; and this, as one after another unexpectedly fell, destroyed them in turn. This certainly renewed the confidence of the fleeing English. Realising the

The finale of the Battle of Hastings. (From It All Happened Before *by John Rador, published by George G Harrap in 1945. Picture by R T Cooper)*

opportunity given to them by the steep bank and by frequent ditches, they unexpectedly halted, pulled themselves together, and inflicted great slaughter on the Normans.'[22]

However, the Duke restored the situation. He counterattacked the English with a strong force and scattered them.

The invader had eventually won a conclusive victory. However, it had been a close-fought conflict and, but for the high-angled arrow-fire tactic, it is possible that the English army might have retained its firm ridge position until nightfall. Thereafter they would have withdrawn to the security of the forest, and then London to build up another, much larger army. The English had actually fought the continental army to a standstill, and it will also be appreciated that half of William's work had been achieved for him in Yorkshire.

Perhaps some 3000 of William's army lay dead or wounded on the field, along with numerous warhorses. Many of these casualties were inflicted in Phases I and II. The English lost a greater number, although no credible calculation of this figure seems to have been made. Most would have involved those tempted forward during the feigned flights. At Hastings a professional army with cavalry and archers eventually just managed to overcome a part-proficient host that comprised predominantly infantry but the English army fought them to a standstill. It is also true that Harold 'lost the battle because his men were unequal to the stress of a purely defensive engagement too long protracted.'[23] However, the Hastings conflict confirmed the effectiveness of Anglo-Saxon/English weapons, which we have studied in this book, especially the various spear forms, the axe and the fine, later swords. The long conflict also endorsed the future potential of well-trained and disciplined footmen fighting defensively. According to Rupert Furneaux, 'The Normans had learned the value of infantry from their near-defeat at Hastings.' Thereafter, the Normans dismounted their armoured troopers at numerous battles to fight, for significant periods, as armoured infantry.

Unknown to William, the major task of England's conquest had already been achieved. Soon, he was crowned King of England. Thereafter, the English were subjected to the rule of a cruel, avaricious tyrant who soon broke his promise not to take English land.

Within three years of his reign one-third of the land had been deliberately laid waste and the populations there were either murdered or died of starvation. Many were relegated to serfdom, being reduced to the status of slave labourers compelled to construct castles. The delicately balanced English democracy admired through much of Europe, so carefully nurtured for centuries and diligently preserved by common law and the strictures of humanitarian and intellectual warrior monarchs such as AEthelstan, disappeared.

Notes and References

Chapter 1
Notes

1. Harold L Peterson, *Daggers & Fighting Knives of the World* (London, Herbert Jenkins).
2. T G E Powell, *The Celts* (London, Thames & Hudson, 1958).
3. Ewart Oakeshott, *The Archaeology of Weapons* (Woodbridge, Boydell Press, reprinted 1994).
4. Oakeshott.
5. Oakeshott.
6. Oakeshott.
7. Oakeshott.
8. Stuart Piggott, 'Swords and Scabbards of the British Early Iron Age', *Proceedings of the Prehistoric Society*, ed. J G D Clark (1950), New Series, vol. XVI.
9. Oakeshott.
10. Piggott.
11. Piggott.
12. Piggott.
13. Piggott.
14 Piggott.

Additional references

Colin B Burgess & Sabine Gerloff, 'The Dirks and Rapiers of Great Britain and Ireland', *Prähistorische Bronzefunde, Abteilung* IV, vol. 7 (Munich, C H Beck'sche Verlagsbuchhandlung, 1985).

Chapter 2
Notes

1. Logan Thompson, 'Roman Roads', *History Today* (February 1997) pp. 21–8.
2. Peter Salway, *Illustrated History of Roman Britain* (London, BCA/Oxford UP, 1993).
3. Keith Branigan, *Roman Britain* (Oxford, Clarendon Press, 1981).
4. Ronald Embleton & Frank Graham, *Hadrian's Wall in the Days of the Romans* (Newcastle on Tyne, Frank Graham, 1984).
5. Branigan.
6. Logan Thompson, *Daggers & Bayonets* (Staplehurst, Kent, Spellmount Publishers, 1999).
7. John Warry, *Warfare in the Classical World* (London, Salamander Books Ltd, 1980).
8. Peter Salway, *Roman Britain* (London, Reader's Digest, 1983).
9. Salway.
10. Salway.
11. Warry.
12. P E Cleator, *Weapons of War* (London, Robert Hale, 1967).
13. Philip Wilkinson, *What the Romans Did For Us* (London, Boxtree, 2000).
14. Wilkinson.

Chapter 3
Notes

1. D M Wilson, *The Anglo-Saxons* (London, Thames & Hudson, 1960).
2. James Campbell, gen. ed., *The Anglo-Saxons* (London, Penguin Books, 1982).
3. H R Ellis Davidson, *The Sword in Anglo-Saxon England* (Woodbridge, Boydell Press, 1962).
4. Ewart Oakeshott, *The Archaeology of Weapons* (Woodbridge, Boydell Press, 1960).
5. Ellis Davidson.
6. Oakeshott.
7. Oakeshott.
8. Frank Stenton, *Anglo-Saxon England* (Oxford, Clarendon Press, 1971, 3rd edn), p. 8.
9. Ellis Davidson.
10. R E M Wheeler, 'London and the Saxons', *Museum of London Catalogue*, No. 6, 1935.
11. John of Antioch, *Fragmenta Historicorum Graecorum*, ed. Didot, Paris, 1874, vol. v, p. 29, in Ellis Davidson.
12. Paul Hill & Logan Thompson, 'The Swords of the Saxon Cemetery at Mitcham', *Surrey Archaeological Collections Journal* (vol. 90, 2003).
13. Janet Lang & Barry Ager, *Swords of the Anglo-Saxon and Viking Periods in the British Museum: A Radiographic Study* (London, British Museum, 1989).
14. Ellis Davidson.
15. Ellis Davidson.
16. Ellis Davidson.
17. R J C Atkinson, *The Construction of the Swords, Scabbards & Shields at the Salisbury & South Wiltshire Museum*, Appendix 1, p.18.

Additional references

Robert Jackson, *Dark Age Britain* (London, Book Club Associates, 1984).
Wilfried Menghin, *Das Schwert im Frühen Mittel-alter, Germanisches National Museum*, (Stuttgart, Konran Theiss Verlag, 1983)
H R Ellis Davidson & E Oakeshott, *Records of the Medieval Sword* (Woodbridge, Boydell Press, 1991).
E Behmer, *Das Zweischneidige der Germanischen Volkerwanderungzeit* (Stockholm, 1939)
R Bruce-Mitford, *The Sutton Hoo Ship Burial* (1978), vol. 2, ch. 6 and pp. 564–82.
H Härke, 'Early Saxon Weapon Burials: Frequencies, Distributions and Weapon Combinations', *Weapons and Warfare*, pp. 49–61

Chapter 4
Notes

1. C W C Oman, *The Art of War in the Middle Ages AD 378–1515* (New York, Great Seal Books, 1953).

2. H V Koch, *Medieval Warfare* (London, Bison Books Ltd, 1978).

3. Ewart Oakeshott, *The Archaeology of Weapons* (Boydell Press, 1994).

4. Oakeshott.

5. Oakeshott.

6. Koch.

7. Agathius, *Historiarum libri quinque, in corsus Scriptorum Historiae Byzantinae* (Bonn, 1828).

8. Oman.

9. Jochen Giesle, 'Frühmittelalterliche Funde aus Niederkassel, Rhein' *Bonner Jahrbucher*, vol. 183. pp. 475–579. Sieg. Kreis (Frankish weapons).

Additional references

Logan Thompson, 'Frankish Throwing Axes in the British Museum Collections'. Department of Prehistory and Europe, British Museum (June 2002).

Chapter 5
Notes

1. Stephen Pollington, *The English Warrior from Earliest Times to 1066* (Norfolk, Anglo-Saxon Books, 1996), p. 116.

2. Michael Swanton, *A Corpus of Pagan Saxon Spear Types*, (Oxford, British Archaeological Reports, 7, 1974).

3. Wheeler, 'London and the Saxons', *Museum of London Catalogue* No. 6, 1935, pp. 169–74.

4. Paul Hill, 'The Nature and Function of Spearheads in England *c.* 700–1100 AD', *Journal of the Arms and Armour Society* (2000), pp. 257–81.

5. Michael Swanton, 'The Spear in Anglo-Saxon Times', PhD thesis, University of Durham, 1966.

6. J Kim Siddorn, *Viking Weapons and Warfare* (Stroud, Tempus Publishing, 2000), pp. 34–5.

7. Siddorn, p. 37.

Chapter 6
Notes

1. Capitulare de Villis, C64, MGH Capitularia, 1, p. 89.

2. Stephen Pollington, *The English Warrior from Earliest times to 1066* (Norfolk, Anglo Saxon Books 1996) pp. 134, 136.

3. Pollington, pp. 134, 136.

4. Pollington, pp. 134, 136.

5. Tania Dickinson & H Härke, *Bosses on Early Anglo-Saxon Shields* (London, The Society of Antiquaries of London, 1992).

6. Paul Hill & Logan Thompson, with contributions by Nick Stoodley: 'Weaponry and Associated Material', *The Early Saxon Cemetery at Park Lane, Croydon, Surrey* by Jacqueline I McKinley, of Wessex Archaeology, *Surrey Archaeological Collections Journal*, 2003.

7. Hill, Thompson & Stoodley.

8. Mortimer Wheeler, 'London and the Saxons', *Museum of London Catalogue No. 6*, 1935.

9. A B Ward Perkins in Wheeler.

10. Wheeler.

11. Janet Lang & Barry Ager, *Swords of the Anglo-Saxon and Viking Periods in the British Museum* (London, British Museum, 1989).

12. D M Wilson, *The Anglo-Saxons* (London, British Museum, 1989), (London, Thames & Hudson, 1960), pp. 110–12, 216–17.

Chapter 7
Notes

1. Magnus Magnusson, *Vikings* (London, Bodley Head, 1980).

2. Magnusson.

3. Magnusson.

4. 'The Genius of the Vikings', Channel 5 television programme (2004).

5. Magnusson.

6. 'The Genius of the Vikings.'

7. James Campbell & Dafydd Kidd, *The Vikings* (London, British Museum Publications Ltd, 1980).

8. Mortimer Wheeler, 'London and the Vikings' *Museum of London Catalogue* (1927).

9. Wheeler.

10. Ewart Oakeshott, *The Archaeology of Weapons* (Woodbridge, Boydell Press, reprinted 1994), pp. 107–118, 131–45.

11. Oakeshott, pp. 107–18.

12. Oakeshott, pp. 107–18.

13. Oakeshott, pp. 107–18.

14. Oakeshott, pp. 88.

15. Oakeshott.

16. Ellis Davidson, *The Sword in Anglo-Saxon England* (Woodbridge, Boydell Press, 1962) p. 52.

Chapter 8
Notes

1. Wheeler, 'London and the Vikings' *Museum of London Catalogue*, (1927)

2. J Campbell & D Kidd, *The Vikings*, pp. 113–17.

Additional references

Holger Arbman, *The Vikings* (London, Thames & Hudson, 1961).

Logan Thompson, 'Stamford Bridge Campaign', *British Army Review*, 1991.

Logan Thompson, 'Hastings Campaign', *Military Modelling*, 1991.

Logan Thompson, 'Fulford & Stamford Bridge Campaign', (unpublished) 2002.

Logan Thompson, 'Stamford Bridge to Hastings: Campaign of 1066', *Army Quarterly & Defence Journal*, vol. 119, No. 3, 1989.

David Howarth, *1066, The Year of Conquest* (London, Collins, 1977).

Chapter 9
Notes

1. Janet Lang & Barry Ager, *Swords of the Anglo-Saxon and Viking Periods in the British Museum: A Radiographic Study* (London, British Museum, 1989).
2. Lang & Ager.
3. Lang & Ager.
4. Lang & Ager.
5. Vera I Evison, 'A Sword from the Thames at Wallingford Bridge', *The Archaeological Journal* (vol. CXXIV, 1967) and *The Royal Archaeological Institute Journal* (1968), pp.160–86.
6. Evison.
7. Evison.
8. Evison.
9. Ellis Davidson.
10. Ellis Davidson.
11. Barry Ager.
12. Barry Ager.
13. Ellis Davidson, *The Sword in Anglo-Saxon England* (Woodbridge, Boydell Press 1962).
14. Ellis Davidson.
15. Ellis Davidson.
16. Ellis Davidson.

Additional references

H R Schubert, *History of the British Iron and Steel Industry* (London, 1957) p. 34.

Peter Bone, 'Development of Anglo-Saxon Swords from the 5th to the 11th Century' *Weapons and Warfare in Anglo-Saxon England*, ed. Sonia Chadwick Hawkes (Oxford University Committee for Archaeology, 1989).

Oxford University Committee for Archaeology, Davidson & Oakeshott, *Records of the Medieval Sword* (Woodbridge, Boydell Press, 1991)

D A Hinton, *The Catalogue of the Anglo-Saxon Ornamental Metalwork, 700–1100,* (Ashmolean Museum, Oxford, 1974).

Chapter 10
Notes

1. H R Ellis Davidson *The Sword in Anglo-Saxon England* (Woodbridge, Boydell Press, 1962), p. 38
2. Ellis Davidson.
3. B J Gilmour 'A Late Anglo-Saxon Sword from Gilling West, N. Yorkshire' *Medieval Archaeology* (vol. XXX, 1986).

4. Gilmour.

Additional references

The Gilling Sword Blade by B J Gilmour

1. R F Tylecote & B Gilmour, 'The Metallography of Early Ferrous Edge Tools and Edged Weapons'
2. Tylecote & Gilmour

Chapter 11
Notes

1. N P Brooks, *Arms Status and Warfare in Late Anglo-Saxon England*, (BAR 59, Oxford, 1978).
2. Brooks.
3. Brooks.
4. Brooks.
5. Brooks.
6. Brooks.
7. Brooks.
8. Stephen Pollington, *The English Warrior from Earliest Times to 1066*. (Norfolk, Anglo-Saxon Books, 1996).
9. David Howarth, *1066, The Year of Conquest* (London, Collins, 1977).
10. Rupert Furneaux, *Conquest 1066* (London, Secker & Warburg, 1966).
11. Furneaux.

Additional references

Snorri Sturluson, *Heimskringla* (the sagas of the Norse kings), vol. IV, pp. 37–8.
David M Wilson, *The Bayeux Tapestry* (London, Thames & Hudson, 1985).
J F C Fuller, *Decisive Battles of the Western World*, (London, Cassell & Co., 1954, 2001), ch. 13, 'The Battle of Hastings'.
William of Poitiers 'The Deeds of William, Duke of the Normans and King of the English', *English Historical Documents*, ed. David Douglas, vol. 11, pp. 217–18.

Chapter 12
Notes

1. C W C Oman, *The Art of War in the Middle Ages AD 378–1515*. (New York, Great Seal Books, 1953).
2. W H Davenport Adams, *Battle Stories from British and European History* (London, Swan Sonnenschein & Co., 4th edn, 1889).
3. Davenport Adams.
4. Rupert Furneaux, *Conquest 1066* (London, Secker & Warburg, 1966).
5. Furneaux.
6. W Spatz, *Schlacht von Hastings* (Berlin, 1886), pp. 33–34.
7. 'The Anglo-Saxon Chronicle (D) 1066', *English Historical Documents*, vol. 2, 1042–1189.
8. J H Round, 'La Bataille de Hastings 1066', *Revue Historique*, vol. 65 (Sept

1897), pp. 61–77.

9. J F C Fuller, *Decisive Battles of the Western World* (London, Cassell & Co., 2001) ch. 13 'The Battle of Hastings', pp. 360–384.

10. *The Chronicle of Florence of Worcester*, translator Thomas Forester (London 1854), p. 164.

11. Furneaux.

12. Furneaux.

13. Fuller.

14. The Anglo-Saxon Chronicle.

15. Furneaux.

16. Furneaux.

17. Spatz.

18. Fuller.

19. William of Poitiers 'The Deeds of William, Duke of the Normans and King of the English', *English Historical Documents*, ed. David Douglas, vol. 11, pp. 217–18.

20. William of Poitiers.

21. Oman.

22. Ordericus Vitalis, *Historia Ecclesiastica Angliae et Normaniae*.

23. F Stenton, *Anglo-Saxon England*, 3rd edn (Oxford, Clarendon Press, 1971), p. 595.

Additional references

N P Brooks, 'Arms Status and Warfare in Late Anglo-Saxon England' (BAR, 122, Oxford, 1978).

Magnus Magnusson, *Vikings* (London, Book Club Associates, 1980).

David Howarth *1066, The Year of Conquest* (London, Collins, 1977).

D M Wilson *The Anglo-Saxons* (London, Thames & Hudson, 1960).

James Campbell, *The Anglo-Saxons* (London, Penguin Books, 1991).

Stephen Pollington, *The English Warrior from Earliest Times to 1066* (Norfolk, Anglo-Saxon Books, 1996).

M K Lawson, *The Battle of Hastings 1066* (Stroud, Tempus, 2002).

Stephen Morillo, *The Battle of Hastings: Sources and Interpretations* (Woodbridge, Boydell & Brewer, 1996).

Jim Bradley, *The Battle of Hastings* (Stroud, Sutton Publishing, 1998).

Index

Page numbers in bold type indicate illustrations.